# INTEGRATED PROCESS DESIGN AND DEVELOPMENT

# THE BUSINESS ONE IRWIN/APICS LIBRARY OF INTEGRATIVE RESOURCE MANAGEMENT

## Customers and Products

Marketing for the Manufacturer   *J. Paul Peter*

Field Service Management: An Integrated Approach to Increasing Customer Satisfaction   *Arthur V. Hill*

Effective Product Design and Development: How to Cut Lead Time and Increase Customer Satisfaction   *Stephen R. Rosenthal*

## Logistics

Integrated Production and Inventory Management: Revitalizing the Manufacturing Enterprise   *Thomas E. Vollmann, William L. Berry and D. Clay Whybark*

Purchasing: Continued Improvement through Integration   *Joseph Carter*

Integrated Distribution Management: Competing on Customer Service, Time and Cost   *Christopher Gopal and Harold Cypress*

## Manufacturing Processes

Integrative Facilities Management   *John M. Burnham*

Integrated Process Design and Development   *Dan L. Shunk*

Integrative Manufacturing: Transforming the Organization through People, Process and Technology   *Scott Flaig*

## Support Functions

Managing Information: How Information Systems Impact Organizational Strategy   *Gordon B. Davis and Thomas R. Hoffman*

Managing Human Resources: Integrating People and Business Strategy   *Lloyd Baird*

Managing for Quality: Integrating Quality and Business Strategy   *Howard Gitlow*

World-Class Accounting and Finance   *Carol J. McNair*

# INTEGRATED PROCESS DESIGN AND DEVELOPMENT

*Dan L. Shunk*

**BUSINESS ONE IRWIN**
Homewood, Illinois 60430

© RICHARD D. IRWIN, INC., 1992

*All rights reserved.* No part of this publication may be reproduced, stored in a retrieval system, or transmitted, in any form or by any means, electronic, mechanical, photocopying, recording, or otherwise, without the prior written permission of the publisher.

This publication is designed to provide accurate and authoritative information in regard to the subject matter covered. It is sold with the understanding that neither the author nor the publisher is engaged in rendering legal, accounting, or other professional service. If legal advice or other expert assistance is required, the services of a competent professional person should be sought.

*From a Declaration of Principles jointly adopted by a Committee of the American Bar Association and a Committee of Publishers.*

Sponsoring editor:   Jeffrey A. Krames
Project editor:   Karen Murphy
Production manager:   Diane Palmer
Designer:   Larry J. Cope
Art coordinator:   Mark Malloy
Compositor:   TCSystems, Inc.
Typeface:   11/13 Times Roman
Printer:   The Book Press, Inc.

### Library of Congress Cataloging-in-Publication Data

Shunk, Dan L.
   Integrated process design and development  /  Dan L. Shunk.
     p.   cm.—(The Business One Irwin/APICS library of integrative resource management)
   Includes index.
   ISBN 1-55623-556-9
   1. Production engineering.   2. Manufacturing processes.
3. Design, Industrial.   I. Title.   II. Series.
TS176.S49   1992
658.5—dc20                                                                                           91-36406

*Printed in the United States of America*

1 2 3 4 5 6 7 8 9 0 BP 9 8 7 6 5 4 3 2

*This book is dedicated to our children, Michael, Benjamin, Carolyn, Kathryn, and Elizabeth, and to all of those children who shall benefit by the progress that this APICS series shall foster in the advancement of a national manufacturing capability.*

# FOREWORD

*Integrated Process Design and Development* is one book in a series that addresses the most critical issue facing manufacturing companies today: integration—the identification and solution of problems that cross organizational and company boundaries—and, perhaps more importantly, the continuous search for ways to solve these problems faster and more effectively! The genesis for the series is the commitment to integration made by the American Production and Inventory Control Society (APICS). I attended several brainstorming sessions a few years ago in which the primary topic of discussion was, "What jobs will exist in manufacturing companies in the future—not at the very top of the enterprise and not at the bottom, but in between?" The prognostications included:

- The absolute number of jobs will decrease, as will the layers of management. Manufacturing organizations will adopt flatter organizational forms with less emphasis on hierarchy and less distinction between white collars and blue collars.
- Functional "silos" will become obsolete. The classical functions of marketing, manufacturing, engineering, finance, and personnel will be less important in defining work. More people will take on "project" work focused on continuous improvement of one kind or another.
- Fundamental restructuring, meaning much more than financial restructuring, will become a way of life in manufacturing enterprises. The primary focal points will be a new market-driven emphasis on creating value with customers, as well as greatly increased flexibility, a new business-driven attack on global markets which includes new deployment of information technology, and fundamentally new jobs.
- Work will become much more integrated. The payoffs will increasingly be made through connections across organizational and company boundaries. Included are customer and vendor partnerships, with an overall focus on improving the value-added chain.

- New measurements that focus on the new strategic directions will be required. Metrics will be developed, similar to the cost of quality metric, that incorporate the most important dimensions of the environment. Similar metrics and semantics will be developed to support the new uses of information technology.
- New "people management" approaches will be developed. Teamwork will be critical to organizational success. Human resource management will become less of a "staff" function and more closely integrated with the basic work.

Many of these prognostications are already a reality. APICS has made the commitment to *leading* the way in all of these change areas. The decision was both courageous and intelligent. There is no future for a professional society not committed to leading-edge education for its members. Based on the Society's past experience with the Certification in Production and Inventory Management (CPIM) program, the natural thrust of APICS was to develop a new certification program focusing on integration. The result, Certification in Integrative Resource Management (CIRM) is a program composed of 13 building block areas which have been combined into four examination modules, as follows:

Customers and products
- Marketing and sales
- Field service
- Product design and development

Manufacturing processes
- Industrial facilities management
- Process design and development
- Manufacturing (production)

Logistics
- Production and inventory control
- Procurement
- Distribution

Support functions
- Total quality management
- Human resources

Finance and accounting
Information systems

As can be seen from this topical list, one objective in the CIRM program is to develop educational breadth. Managers increasingly *must* know the underlying basics in each area of the business: who are the people who work there, what are day-to-day *and* strategic problems, what is state-of-the-art practice, what are the expected improvement areas, and what is happening with technology? This basic breadth of knowledge is an absolute prerequisite to understanding the potential linkages and joint improvements.

But it is the linkages, relationships, and integration that are even more important. Each examination devotes approximately 40 percent of the questions to the connections *among* the 13 building block areas. In fact, after a candidate has successfully completed the four examination modules, he or she must take a fifth examination (Integrated Enterprise Management), which focuses solely on the interrelationships among all functional areas of an enterprise.

The CIRM program has been the most exciting activity on which I have worked in a professional organization. Increasingly, manufacturing companies face the alternative of either proactive restructuring to deal with today's competitive realities, or just sliding away—giving up market share and industry leadership. Education must play a key role in making the necessary changes. People working in manufacturing companies need to learn many new things and "unlearn" many old ones.

There were very limited educational materials available to support CIRM. There were textbooks in which basic concepts were covered and bits and pieces which dealt with integration, but there simply was no coordinated set of materials available for this program. That has been the job of the CIRM series authors, and it has been my distinct pleasure as series editor to help develop the ideas and facilitate our joint learning. All of us have learned a great deal, and I am delighted with every book in the series.

**Thomas E. Vollmann**
**Series Editor**

# PREFACE

The world of process design and development has changed immensely over the past 10 to 20 years. Originally, in the serial fashion of design and manufacturing, *process* design and development was considered after the *product* was developed. But in today's integrated world, product design and process design are developed in a simultaneous, integrated fashion. In an attempt to understand how to develop integrated resource management (IRM) for a successful 21st century company, the role of process design and development must be well understood. In this book we shall attempt to outline the notions that exist today as they relate to the product-process life cycle and the definitions of terminology and the changing terminology.

To clarify these ideas we will provide a few examples. This book will then go through a rigorous process including a look at the fundamental building blocks for integrated processes, the planning needed, the control of the processes, and the execution of an integrated process. This is followed by business measurements that integrated processes must achieve and concludes with the information requirements and the trends and directions that process design shall take in the 21st century.

The book uses three case study examples to demonstrate the concepts. The first is a simple case study example of how a part family at Rockwell International was analyzed and integrated into a very efficient production cell with order-of-magnitude reductions in throughput times. It will also compare the methods of how a part is made in three different environments: a metallic environment, an electronics environment, and a nonmetallic, or composites, environment. We shall then migrate to a more detailed case study example of how an American producer of world-class precision gear products is attempting to develop an integrated product-process-assembled product environment for a specific product family that uses the principles of process design and development described in this text. Finally, a rather radical, yet totally integrated, product-process case study example from Japan is examined in which the company has developed a new paradigm for product and process development in a simultaneous, or concurrent, environment that truly exploits not only the technology but also the people aspects of integration.

**Dan L. Shunk**

# ACKNOWLEDGMENTS

The creation of this book was through the vision of APICS and its members to create a certification process for the professional associated with integrated resource management. I applaud this visionary body, and I appreciate their confidence in my ability to take a tough topic such as integrated process design and development and translate this into a text that is acceptable to the APICS member.

I also want to express my appreciation for a key APICS member who had the strength of conviction in our friendship to say, "Sure, Dan, go ahead and write it." To Jim Breslauer, I offer that we are still friends—good friends. You are the best editor in the world.

To my professional colleagues who reviewed the text, I also wish to express my sincere appreciation. This includes Jim, Roger Willis, and Marlon Ritchie. Thanks for the invaluable content insight.

Finally, to my family of helpers, I truly appreciate your technical as well as your emotional support. This includes my wife, Gail, and Jan Murra. We make a good team.

**D.L.S.**

# CONTENTS

**CHAPTER 1**    **THE PRODUCT-PROCESS LIFE CYCLE FOR INTEGRATED PROCESS DESIGN AND DEVELOPMENT**    **1**

Section 1.1 The Objectives of Integrated Process Design and Development, 1

Section 1.2 The Linkages of Process with Product Design, 3

Section 1.3 Strategic Implications of Integrated Product-Process Design and Development, 4

Section 1.4 Linking Process Design with Customer Needs and Expectations, 6

Section 1.5 Establishing Corporate Metrics for Product-Process, 9

Section 1.6 The Strategic Planning of Focused Factories, 9

Section 1.7 Linking the Process Needs with Capabilities, 11

Section 1.8 An Overview of the Trends in Process Design, 11

Section 1.9 Addressing the Obstacles to Integration, 14

Section 1.10 An Overview of the Integrated Process Design Book, 15

**CHAPTER 2**    **INTRODUCTION TO BASIC PROCESSES AND BASIC TERMINOLOGY**    **19**

Section 2.1 Introduction to Basic Processes, 19

Section 2.2 Basic Terminology, 20

Section 2.3 Basic Information Terminology Necessary for Product Design, 22

Section 2.4 The Metrics—Definitions for Process Design, 25

Section 2.5 Introduction to Types of Manufacturing, 29

**CHAPTER 3**    **INTRODUCTION TO BASIC PROCESS DESIGN USING THREE EXAMPLES**    **31**

Section 3.1 An Example of a Metallic Part Process Design and Development, 32

Section 3.2 An Example of How a Nonmetallic Part Is Produced, 39

xv

Section 3.3 An Example of How an Electronic Part Is Produced, 41

CHAPTER 4  PREREQUISITES FOR INTEGRATED PROCESS DESIGN AND DEVELOPMENT  43

Section 4.1 Concurrent Engineering, 44
Section 4.2 The Concept of Group Technology, 51
Section 4.3 Total Quality Management, 56
Section 4.4 Material Management Implications for Integrated Process Design, 57
Section 4.5 Process Capability Design, 60
Section 4.6 Benchmarking, 61
Section 4.7 Material Handling and Material Flow System Implications for Integrated Process Design, 63
Section 4.8 Setup and Tooling Strategies Including "Single Minute Exchange of Dies," 65
Section 4.9 System Integration Design Philosophy, 69
Section 4.10 Introduction to an Integrated Process Design and Development Systems Definition Methodology, 70
Section 4.11 Simulation and Its Role in Integrated Process Design and Development, 77

CHAPTER 5  PLANNING THE PROCESS FOR INTEGRATED PROCESS DESIGN AND DEVELOPMENT  85

Section 5.1 Performing a Technology Assessment, 85
Section 5.2 Benchmarking the Integrated Process, 87
Section 5.3 Achieving Flexibility Due to Product Definition Uncertainties, 90
Section 5.4 Defining the Process Needs for a Product, 92
Section 5.5 Creation of Various Types of Prototypes, 94
Section 5.6 Creation of the Process Plan, 97
Section 5.7 Computer-Aided Process Planning, 103
Section 5.8 Performing the Make-Buy Decision, 105
Section 5.9 First Item Planning with CPM, PERT, and Simulation, 106
Section 5.10 The Infusion of Total Quality Management Principles into Process Design, 108

CHAPTER 6  CONTROLLING THE PROCESS  117

Section 6.1 Introduction to Control, 117
Section 6.2 History of Automatic Control, 120

Section 6.3 Examples of Modern Control Systems, 121
Section 6.4 Controlling the Flow of Product through the Factory, 123
Section 6.5 The History of Numerical Control, 125
Section 6.6 Machine Control Concepts, 126
Section 6.7 Principles of Operation, 129
Section 6.8 Types of Control Systems, 133
Section 6.9 The Creation of the Geometry Profile for the Process, 134
Section 6.10 Advanced Numerical Control Developments, 136
Section 6.11 Closed-Loop Control, 138
Section 6.12 Controlling the Inventory and Material Flow, 139
Section 6.13 The Migration to Integrated Processes, 140

CHAPTER 7 PROCESS METRICS 145

Section 7.1 Traditional Metrics, 145
Section 7.2 Changing Strategic Directions, 146
Section 7.3 The New Metrics for World-Class Competition, 148
Section 7.4 Measuring the Efficiency and Effectiveness of Manufacturing, 152
Section 7.5 The Learning Curve Notion, 154

CHAPTER 8 GETTING STARTED IN INTEGRATED PROCESS DESIGN AND DEVELOPMENT (A NATIONAL SURVEY AND TWO CASE STUDIES) 157

Section 8.1 Summary Results of the National Science Foundation Survey for Integrated Product Design in Manufacturing, 163
Section 8.2 A Case Study of a World-Class Manufacturer and the Integrated Approach to Competitive Designs, 167
Section 8.3 The Status of Concurrent Engineering Implementation at the Garrett Engine Division of Allied Signal Corporation, 173

CHAPTER 9 INFORMATION SYSTEM REQUIREMENTS AND DESIGN FOR INTEGRATED PROCESS DESIGN AND DEVELOPMENT 181

Section 9.1 Process Design and Development Information Systems—An Overview, 184

Section 9.2 An Architecture for Integrated Process Design and Development, 187

Section 9.3 Definition of Computer Integrated Process Requirements, 190

Section 9.4 Process Control Information Systems, 193

Section 9.5 Design of the Integrated Management and Control Information System, 194

Section 9.6 The Definition of Modern, Integrated Process Execution Information Requirements, 197

Section 9.7 Requirements for Financial Systems Integration, 203

Section 9.8 Legacy Systems and the Need for Dynamically Migratable Systems, 206

Section 9.9 Proactive Systems Designed for Proactive Management, 208

CHAPTER 10   DIRECTIONS AND TRENDS   211

Section 10.1 Concurrent Engineering, 213

Section 10.2 Total Quality Management Directions and Trends, 218

Section 10.3 Material Supplier Integration, 219

Section 10.4 Equipment Supplier Integration, 221

Section 10.5 Process Planning Directions and Trends, 222

Section 10.6 Process Control Directions and Trends, 223

Section 10.7 Asset Management and Capacity Planning in the Integrated Process Design, 224

Section 10.8 Rapid Prototyping, 225

Section 10.9 Flexible, Focused Factories for Integrated Product-Process Design, 226

CHAPTER 11   AN INTEGRATED PRODUCT-PROCESS CASE EXAMPLE   229

Section 11.1 An Assessment of Integration in Japan, 233

Section 11.2 The Nippondenzo Case Study, 233

Section 11.3 New Metrics for This Integrated Product-Process Example, 237

Section 11.4 The Role of the Process Design Function, 237

CHAPTER 12   PRIMARY INTERFUNCTIONAL CONNECTIONS FOR PROCESS DESIGN AND DEVELOPMENT   239

Section 12.1 Linkages for Process Design and Development, 239

Section 12.2 The Cultural Implications of Integrated Process Design and Development, 244

Section 12.3 The Effects of Integrated Process Design and Development on Sally, Mary, Dan, and Bill, 247

**OTHER TITLES IN THE BUSINESS ONE IRWIN/APICS LIBRARY OF INTEGRATIVE RESOURCE MANAGEMENT**     251

**INDEX**     253

# INTEGRATED PROCESS DESIGN AND DEVELOPMENT

# CHAPTER 1
# THE PRODUCT-PROCESS LIFE CYCLE FOR INTEGRATED PROCESS DESIGN AND DEVELOPMENT

## SECTION 1.1 THE OBJECTIVES OF INTEGRATED PROCESS DESIGN AND DEVELOPMENT

In the vision for the factory of the 21st century, several things become evident. As Drucker states in his *Harvard Business Review* article, "We cannot build it yet, but already we can begin to specify the post-modern factory of 1999."[1] The first of four key elements of that factory will be that *statistical quality control* will exist throughout the entire product-process life cycle scenario, not just on the manufacturing floor. Therefore, it is necessary to understand what accountability is required and to align the information system requirements with that accountability.

A second key is that *new manufacturing accounting systems* must be in place that will allow managers to truly track value-added and non–value-added contributions from the product and the process. For this, new metrics and new measurement techniques will stray from the traditional measures of a direct labor basis to actually identify the cost of nonproduction.

A third key will be *modular manufacturing*. Most large companies are migrating to global environments for a myriad of reasons—some for strategic alliances, others simply for offsets. This tendency has forced the realization that the world's political borders are becoming more transparent to the manufacturing environment. Modular manufacturing will be made up of flotillas of integrated modules: integrated design modules, integrated manufacturing modules, integrated service modules, integrated materials management modules, and so on. Each of these modules may be spread throughout various locations on the globe.

Finally, to make this global transition, managers must understand the *systems approach* and how one can operate in this fully integrated, fully global basis. The bottom line of Drucker's article is that to compete in the 21st century, costs must be in line with the competition, and quality must be at the "six sigma" level of world-class performance (a notion that will be developed and refined in Chapter 7). However, *time* will be the major new metric by which most world-class manufacturers will be judged. This is Drucker's conclusion and this author agrees.

To do this, Mr. Bob Galvin, chairman of the board of Motorola, has described the following as his vision for where enterprise integration is going. He states, "The nation, and parenthetically the company, that masters the management of information for the service industries, along with owning the key parts of the service sector, is destined for global economic leadership of historic proportions."[2]

This requires two fundamental paradigm shifts for the enterprise:

1. That the enterprise is a service that uses manufacturing as the basis by which wealth is generated.
2. That the management of information in this service environment can provide a major global competitive edge to the company that can achieve the optimal integration of that information with technology, with the experience and knowledge of the people, and with the full communication links between the enterprise and its customers and suppliers.

To effect this change, significant advancements have happened in the design and manufacture of products. Fundamental shifts are occurring with the integration of the design and drafting functions, with the planning and scheduling functions, with the purchasing and materials management functions, and with the fabrication, assembly, and test functions to ultimately create the integrated enterprise. This enterprise engulfs not only the firm as we know it today, but also begins with the customer, integrates with the distribution systems and the supplier systems, and ultimately returns to the customer with a product that meets the customer's needs.

To summarize, the fundamental transition for the enterprise of the 21st century can be captured in three distinct elements. The first element is that the employees must shift from single-discipline employees to those with multiple-discipline capabilities. The second is that the equipment must shift as well from stand-alone equipment to integrated equipment and systems. To accommodate these shifts, the third element is that the process design and development capabilities of the enterprise must be integrated into the overall life cycle of the product. This book will develop all three elements.

## SECTION 1.2  THE LINKAGES OF PROCESS WITH PRODUCT DESIGN

In the past, the product was designed and then the process was developed to provide the resources needed to produce the product. In this serial fashion, often a "wall" developed between the way the product was designed and the process was developed. But over the past 10 to 20 years, manufacturers have demonstrated that significant savings of time, as well as significant improvements in quality and in overall product cost, can be achieved by lowering this wall and eliminating the barriers between design and manufacturing. The trend toward concurrent engineering (or integrated product-process development, IPPD) has come to represent the notion that during the course of the product-process development life cycle, the team of design engineers, manufacturing engineers, purchasing agents, systems integrators, customer representatives, supplier representatives, and many others must get together in some productive fashion to develop the product-process characterization.

If one analyzes the traditional life cycle of a product through the stages of conceptual, preliminary, and detail design; advanced development; prototyping; fabrication; assembly; and then into the customer's hands for operations and support and finally retirement, it becomes imperative to understand how the costs are actually determined. Dr. Eugene Merchant showed as early as 1970 that at the end of conceptual design, normally only 1 to 5 percent of the total life cycle cost has been expended. However, at the end of conceptual design, based on recent National Science Foundation surveys conducted at Arizona State University, the enterprise has committed up to 80 percent of the life cycle cost. This is shown graphically in Figure 1–1. Thus, the integration of process design and development early in the conceptual and preliminary design stages is the cornerstone of the overall, integrated product-process development cycle.

## SECTION 1.3  STRATEGIC IMPLICATIONS OF INTEGRATED PRODUCT-PROCESS DESIGN AND DEVELOPMENT

Given the data stated previously, the recognition and the responsibilities of developing a product and process in an integrated environment are shifting. The acquisition process for all subcomponents requires that the supplier base be linked to the overall process development cycle. Research and development investments now are being made not only in terms of the product but also in the integrated set of tools necessary to develop the product-process environment. The equipment and facilities must also be integrated into the plan.

To this end, many companies have recognized that the integra-

**FIGURE 1–1**
**Life Cycle Costs**

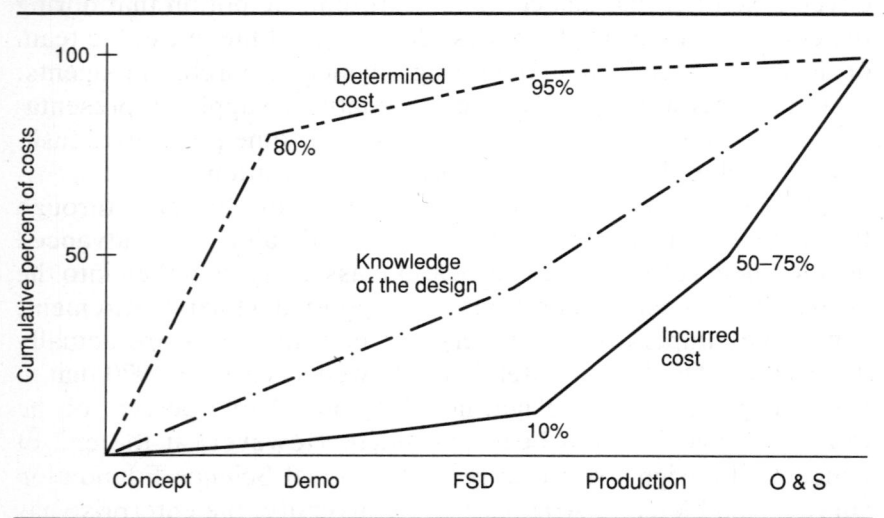

tion of product and process life cycle provides a sustainable competitive advantage over the competition from the viewpoint of timeliness in servicing the customer. Companies are realizing that certain core competencies are fundamental to the success of that organization.

Core competencies is a new strategic notion developed by Prahalad and Hamel in their *Harvard Business Review* article of 1990.[3] In this article, they note how NEC was able to view its corporation much differently than its competitors because it looked at the assets of the corporation as its "competencies" for creating and performing world-class tasks. These include the ability to truly understand what the customer needs, to design world-class semiconductor circuits, or to service the customer around the world better than anyone else. Unlike the typical physical assets of today, core competencies do not (at least should not) deteriorate. The authors then portray how these core competencies can be utilized to create products and services.

This is a paradigm shift that must be understood as the strategic planning processes of the 1990s take shape. As one of the perceived cornerstones to the planning processes, the notion of core competencies often is intuitively counter to a rapid return on investment. For example, Boeing, a successful company with a huge backlog of work, has mastered two outwardly visible core competencies that set it apart. One of these is called "AOG" for "airplane on the ground." Through the AOG focus, Boeing has developed a service policy for all its customers that ensures a spare part will be in the air to fix a downed aircraft in a matter of hours once the problem is identified. To do this, Boeing has strategically invested in a huge memory bank used to store the configuration of all Boeing aircraft in operation. Outwardly, this huge memory bank cannot have a superior return on investment over some process improvement tools needed for today's production. However, the customer wants the airplane to fly, and for this to happen, Boeing must support the customer with spare parts. A tactical, economics view of the situation says Boeing shouldn't invest, but the strategic customer view says this core competency sets Boeing apart and has aided it in capturing billions of dollars of backlog.

## SECTION 1.4 LINKING PROCESS DESIGN WITH CUSTOMER NEEDS AND EXPECTATIONS

Manufacturing thinking now recognizes that customer requirements and customer success are the major benchmarks by which companies can measure their overall effectiveness. Manufacturers once viewed financial return to shareholders as an independent variable that could be directly managed. The shift to the belief that financial return to shareholders is a dependent variable that is the result of overall customer satisfaction has led to major new methodologies to meet customer desires. Table 1–1, devised by a senior executive of a major multinational corporation, attempts to demonstrate how the shifting manufacturing paradigms affect corporate strategy and the design of integrated systems.

What Table 1–1 represents is reflected in this text. The closed-loop, integrated product-process design and development environment must be understood in the context of the customer and the employee for it to be successful.

This text shall follow the simple methodology shown in Figure 1–2 as a basis for linking customer satisfaction with business strategy and with overall technological strategy.

Specific, strategic questions are asked at each level of this four-step process to more fully understand the customer and the resources of the organization needed to meet the customer requirements, as shown in Figure 1–3.

The questions are very simple to ask, yet very important to answer with the full understanding of the organization.

- Who are the customers?
- Who are the competitors?

**TABLE 1–1**
**A Comparison of Views on the Focus of the Business**

| From | To |
|---|---|
| 1. Shareholder | 1. Customer |
| 2. Employee | 2. Employee |
| 3. Customer | 3. Shareholder |

**FIGURE 1–2**
**Linking the Customer to the Business to the Technology**

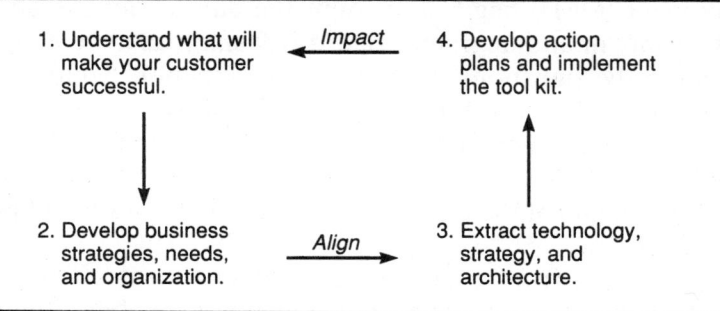

1. Understand what will make your customer successful. — *Impact* → 4. Develop action plans and implement the tool kit.

2. Develop business strategies, needs, and organization. — *Align* → 3. Extract technology, strategy, and architecture.

**FIGURE 1–3**
**The Strategic Questions for World-Class Manufacturing**

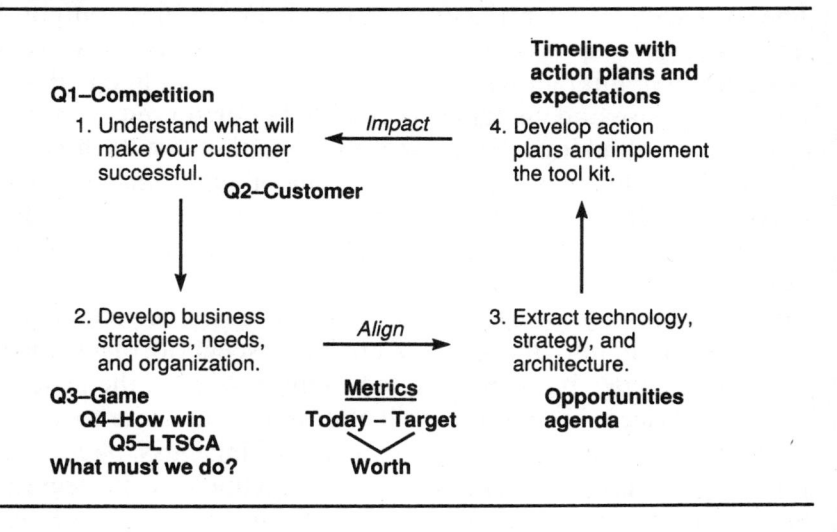

Q1–Competition
1. Understand what will make your customer successful. — *Impact* → 4. Develop action plans and implement the tool kit.
   Timelines with action plans and expectations

Q2–Customer

2. Develop business strategies, needs, and organization. — *Align* → 3. Extract technology, strategy, and architecture.

Q3–Game
Q4–How win
Q5–LTSCA
What must we do?

Metrics
Today – Target
Worth

Opportunities agenda

- What competencies are required?
- What game are we playing?
- How do we win?

Questions such as these force the organization to look at the total enterprise for success. The notion of long-term, sustainable competitive advantages (LTSCA) was introduced by Wheelwright and others in the 1980s to focus corporate strategic planning on these very questions. Formal methodologies such as quality function deployment (QFD) begin to present the data about the customer needs/wants in an integrated, rigorous fashion as a stepping-stone to answer the questions above. Quality function deployment involves the customer in the overall design and analysis methodologies, and they are included in the formal planning processes, where formal methodologies are used to define how the characteristics and the requirements of the customer compare with the overall product characteristics.

The results from quality function deployment often are startling. They are much broader in concept and capability than originally conceived simply with an internal design team. Also, the packaging of the product often is significantly improved because that is the overt integration between the product and the customer. The packaging of the product will provide better input-output, better customer response, or whatever the metric is by which the customer will interact with a product. The bottom line is that customer response and customer needs are better served.

Modern thinking is based on the recognition that the first step in developing a business strategy is to understand what is going to make the customer successful. This requires a true understanding of who the customer's customer is. Step two takes that customer requirement and/or perceived requirement, and that set of customer metrics by which we can achieve customer success, and translates that into the overall business scenario. The business scenario includes but is not limited to the products, the processes, and the services the company will offer to the customer to make the customer successful. Aligned with these business plans is step three, which is the development of the technology strategies, the marketing strategies, and the human resource strategies necessary to develop the actions in step four that will ultimately affect the customer. This four-step process is fundamental to the success of the overall enterprise.

To succeed at this new product-process life cycle endeavor, one must recognize that there are various ways to develop products. Many companies believe prototypes are necessary to develop products. As we develop the notion of product-process life cycle, a complete spectrum of prototyping is required. If no advancement in product or process technology is required for a new offering, simply an enhancement of it, then no prototyping will be necessary. That's to the left of the spectrum. To the right of the spectrum is where new product notions are attempted that will be coupled with new process notions. Here, characterization of both product and process require extensive prototyping. As one views the spectrum from no prototyping to extensive prototyping, many, many scenarios can be played out. This shall be developed in detail in Section 5.5 of this book.

## SECTION 1.5   ESTABLISHING THE CORPORATE METRICS FOR PRODUCT-PROCESS

People tend to respond to the way they are judged, measured, and rewarded. The new corporate strategies of the 21st century are changing the way people and the product-process life cycle are measured. In Chapter 7, we shall develop and analyze what the new process life cycle metrics are to be successful in the 21st century manufacturing environment. In this chapter, the notion of "time-based competition" as developed by George Stalk[4] is coupled with the "core competencies" notion of Prahalad and the strategic notions of strategic alliances and continuous improvement to form a set of four criteria that are forcing change in the corporate metrics.

## SECTION 1.6   THE STRATEGIC PLANNING OF FOCUSED FACTORIES

Dr. Wickham Skinner introduced the notion of focused factories in the 1960s.[5] In his landmark paper, he introduced the concept that the factory must be "focused" by being very efficient and very flexible at the same time. Many companies are going to the focused-factory concept as a means of overcoming the existing "legacy systems" burden of current information systems and cultures. These companies do not wish to burden the newer concepts with old products and procedures. This allows the new center to test the idea of a core competency before introducing the notion to the entire factory.

To achieve this efficiency and flexibility, the product and the process must be fully integrated, and certain key metrics such as an economic order quantity of one (EOQ = 1) must be introduced into the factory environment. An EOQ of one requires that setup times and changeover times from one product to another, from one process to another, must approach zero. This is consistent with the view that Dr. Steven Wheelwright took as he attempted to define manufacturing competitiveness in the 21st century.[6] He defines the environment for this competitive endeavor to be one of global intensity and rapid change. The business base will be built on a global environment. Rapid change will be necessary for everything from responding to customer demands to installing an information system. "The minimum requirement to be competitive in the 21st century will be one of operational excellence as perceived today," he writes.

Hence, six sigma in its broadest form (where six sigma is the difficult notion of having no more than 3.4 errors per million opportunies) will be applied not only to the product, for its characterization in meeting the customer requirements, but also to the process, for full replication of the environment to produce a consistent quality product every time. Six sigma will also be applied to the notions of data, time, talent base, and so on. These expanded notions will be developed later in this text. The manufacturing task for the product-process life cycle will be one of integration, interaction, and a term that Dr. Wheelwright introduces called *complementarity,* where design must complement manufacturing, manufacturing must complement service, service must complement the customer, and so forth.

Finally, in the 21st century, manufacturing will exhibit increasing flexibility, efficiency, and effectiveness at the same time. To do so, the enterprise must provide a distinctive competence based on the knowledge that people are the top asset and that continuous learning and improvement are fundamental to the success of the enterprise.

The attempt here is to recognize that the customer is fundamental to the success of any enterprise. The customer wants the manufacturing environment to appear as though it's right across the street, such that the supplier provides the customer with a product and service on as timely a basis as possible. Yet in order to accomplish and accommodate economies of scale as well as economies of scope,[7] we must have globalization of our strategies. This book introduces the idea that "globalization with localization" will be a fundamental building block to successfully competing in the 21st century. The prerequisites for globalization with localization are outlined in Chapter 4. Essentially, the domain of product and process design and development must take on a worldwide capability with communication and standardization of the design and development processes key to overall success.

## SECTION 1.7  LINKING THE PROCESS NEEDS WITH CAPABILITIES

Translating our requirements to be very responsive to our customers and to maintain excellent quality and cost into systems provides an excellent yardstick to measure our system requirements. For example, today most routings found in MRP systems are fixed. This is for planning purposes and probably gives adequate planning visibility. But as we move from planning to control and execution, flexibility becomes more important. We must understand how to achieve flexibility in the routings of the parts by allowing alternative routings or even by transmitting the process needs to a supplier that has a different set of process capabilities. This paradigm shift to flexibility in routings reflected in the detailed process plans shall be addressed in Chapters 6, 9, and 10.

## SECTION 1.8  AN OVERVIEW OF THE TRENDS IN PROCESS DESIGN

Certainly the trends of process design must reinforce the trends appearing in the enterprise in general. Dr. Charles Savage in his text on *Fifth Generation Management*[8] has recognized that when the term *CIM* was coined from the advanced notion of computer integrated manufacturing by Dr. Joseph Harrington, Jr., in 1973,[9] the primary focus was looking at the enterprise from the firm's viewpoint where design and manufacturing were attempted to be integrated with sales and support. However, in Savage's text, published in 1990, the enterprise has a much bolder, much broader, more

## FIGURE 1–4a
## The Enterprise View from Dr. Joseph Harrington, Jr.

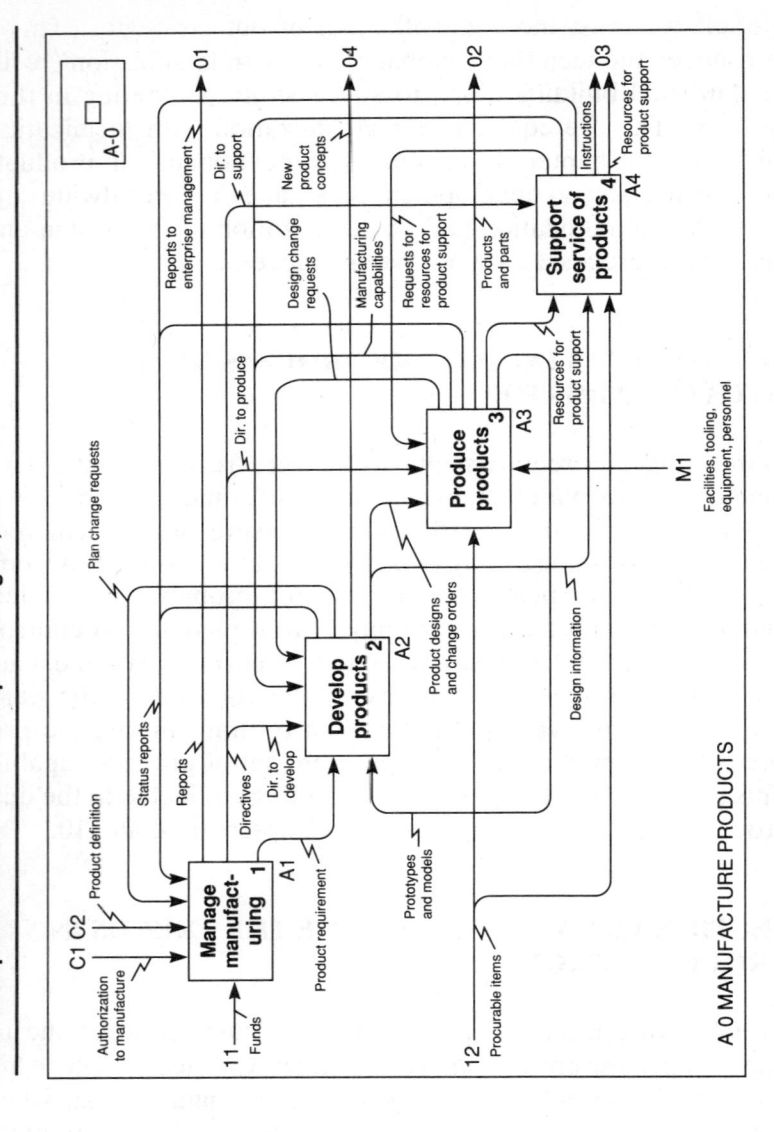

A 0 MANUFACTURE PRODUCTS

grandiose vision. We must begin by recognizing not only the firm but also the integration of the firm with the supplier base, the distribution channels, and the customer to understand what the customer requirements are. Figure 1–4a shows Dr. Harrington's view of the enterprise in 1973, while Figure 1–4b develops the same notion as proposed by Dr. Savage in 1990. Please note the extreme change in scope of these two views.

Several key elements will characterize this enterprise of the 21st century.

1. It will be characterization-driven, by this we mean we understand the product-process and each element of the product and process, and understand the physics and the chemistry or whatever other basic foundations on which we will build the overall effort.
2. The computing environment will be seamless in order to support the information systems required to develop this characterization of product and process.
3. New tools, such as neural networks and others, will supplement the corporation's number-one asset, its knowledgeable work force in the development of new ideas, new notions, and new abilities.

**FIGURE 1–4b**
**The Enterprise View from Dr. Charles Savage**

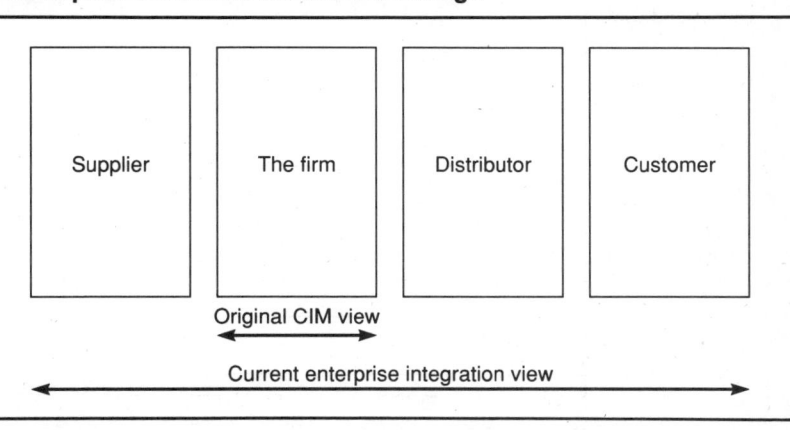

| Supplier | The firm | Distributor | Customer |

Original CIM view

Current enterprise integration view

Source: Charles Savage, *Fifth Generation Management* (Boston: Digital Press, 1989).

4. The focus will be at the enterprise level as compared to simply at the firm level.

The fundamental transition will be in the minds of the work force as we move from single-disciplined employees to multidisciplined employees that have a much broader view of the enterprise and a much better understanding of how to make their customers successful.

## SECTION 1.9  ADDRESSING THE OBSTACLES TO INTEGRATION

This book does not address "rocket science"! The concepts to achieve integrated product and process design and development require few formulas and few technological advancements. Then what is keeping this integration from succeeding? A 1990 survey of industry executives conducted by *Industry Week* revealed the obstacles illustrated in Figure 1–5.

**FIGURE 1–5**
**The Primary Obstacles to CIM Technologies in the United States**

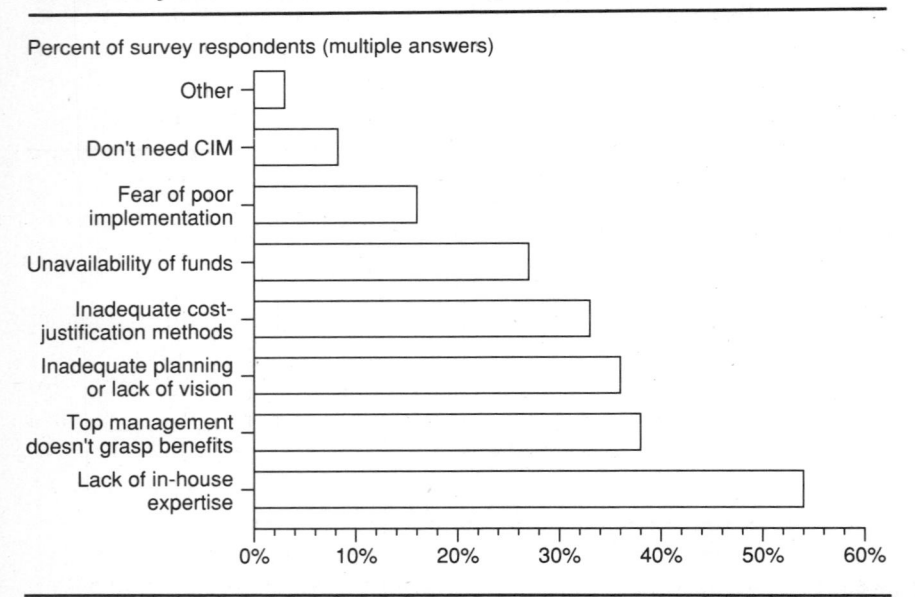

Percent of survey respondents (multiple answers)

| Obstacle | Percent |
|---|---|
| Other | ~3% |
| Don't need CIM | ~8% |
| Fear of poor implementation | ~16% |
| Unavailability of funds | ~27% |
| Inadequate cost-justification methods | ~33% |
| Inadequate planning or lack of vision | ~36% |
| Top management doesn't grasp benefits | ~38% |
| Lack of in-house expertise | ~54% |

Source: 1990 *Industry Week* survey.

The top three reasons in Figure 1-5 are not that the solutions are not perceived to be available. The most-often-cited reason is "lack of in-house expertise." U.S. companies need to find the knowledge necessary to develop the confidence that success can be achieved. The second reason is "top management doesn't grasp the benefits." The significant shifts in the paradigms presented earlier are causing management to rethink what the real benefits must be to be successful against global competition. This benefits assessment must view the enterprise, not just the production function. We will assess the *strategic* investments of companies, such as Boeing, that make large investments in customer support because their core competencies of service dictate this. Finally, the third reason is "inadequate planning or lack of vision." One of the purposes of this text is to create the vision and to outline the planning and control necessary for success.

## SECTION 1.10   AN OVERVIEW OF THE INTEGRATED PROCESS DESIGN BOOK

The world of process design is changing immensely and will continue to change over the next 10 to 20 years. This text is dedicated to an assessment of where the product-process design is as it fits into the overall product-process life cycle, and then it addresses some of the key issues as they appear on the horizon on the 21st century. The book begins by looking at the product-process life cycle and the strategic ramifications of where the product-process life cycle fits. The book then develops some basic terminology for how process information and other types of manufacturing terminologies will play and how they fit together.

Chapter 3 develops one relatively simple case study example for a simple mechanical part to ensure that the reader is well grounded on the notions of how metal parts are produced. The chapter also includes introductions to nonmetallic products and electronic products and displays the similarities and differences in how these integrated processes are designed and developed in today's standards-based environment.

In Chapter 4, the design of the process will be developed. We begin by understanding how the capability and the concept of group technology can be employed and the impact this can have in the product-process overall life cycle. Key issues such as material management and how just-in-time philosophy dovetails with process design are explored along with such issues as systems integration, material handling, environmental planning, and tooling. All aspects necessary to plan the production of the product will be developed to support the planning of the process needs developed in Chapter 5.

Chapter 5 relates to the planning of the process. We begin by looking at the technologies of capability assessment and development of both a variant and a generative planning methodology to capture and characterize this. We also look at various methodologies for prototyping the process to ensure that we understand the complete product-process development cycle.

Chapter 6 develops how the control for the process will be conducted at the factory level down to the machine level. The history of numerical control (NC) shows how technology began to change in the early 1950s and has continued with remarkable speed to date. Such concepts as NC, cutter location file, computer-aided design (CAD), and computer-aided manufacturing (CAM) integration are developed.

Chapter 7 presents a methodology and an assessment of whether the traditional metrics by which manufacturing is judged is adequate for today's customer-oriented, competitive environment. Some of the modern metrics are introduced—those that are representative of where the product-process life cycle must go into the 21st century. In Chapter 7, a key notion is the recognition that the learning curve is changing. Chapter 7 also addresses continuous improvement and how this "inching rapidly"[10] idea espoused by Toyota and other Japanese manufacturers can be included in the overall product-process development cycle.

Chapter 8 develops a more detailed, total life cycle case study for a precision assembled product. This includes the conceptual and preliminary design considerations that affect the product-process, the creation of the product geometry, the creation of the process plan, the detailing of the plan, the tool design, the preparation for execution, and the ultimate requirement for a closed-loop process control system.

Chapter 9 captures what information is going to be required, details some of the necessary data-collection methodologies, and outlines an integrated process development information base scenario. It also looks at the overall integrated information system to support this in a computer integrated process design model, and it takes a distinct departure from some of the rigid routing and process planning information structures of today to look into the integrated information system requirements of the 21st century.

Chapter 10 assesses the directions and trends this integrated process design and development environment is taking. The notions of concurrent engineering, total quality management, and materials supplier integration are but a few of the advanced concepts that must be mastered for integrated process design and development to succeed. These notions include many of the modern issues, such as rapid prototyping, and will be developed to the point where we understand where they fit into the overall directions and trends of modern materials management and modern process design.

Chapter 11 presents an integrated product-process case study example where we review the ideas of Nippondenso, a Japanese company, and some of the shifting paradigms it is using to take a new strategic new-product development look, where product and process are developed simultaneously, where new metrics are in place, and where significant success has been achieved.

The book concludes with the primary interfunctional connections necessary to understand how the enterprise of the 21st century must operate from a people perspective.

## REFERENCES AND RECOMMENDED READINGS

1. Peter F. Drucker, "The Emerging Theory of Manufacturing," *Harvard Business Review*, May–June 1990, pp. 94–102.
2. Robert Galvin, Jr., "Motorola Enterprise Integration Committee," Motorola Company Report, Schaumburg, Ill., 1990.
3. C. K. Prahalad and G. Hamel, "The Core Competence of the Corporation," *Harvard Business Review*, May–June 1990.
4. George Stalk, Jr., "Time—The Next Source of Competitive Advantage," *Harvard Business Review*, July–August 1988, pp. 41–51.
5. Wickham Skinner, "Manufacturing—Missing Link in Corporate Strategy," *Harvard Business Review*, May–June 1969, pp. 136–145.

6. Steven Wheelwright, "Restoring the Competitive Edge in U.S. Manufacturing," *California Management Review*, 1985.
7. Joel Goldhar and Mariann Jelinek, "Manufacturing as a Service Business," *Computers in Industry*, 1990.
8. Charles Savage, *Fifth Generation Management* (Boston: Digital Press, 1990).
9. Joseph Harrington, Jr., *Computer Integrated Manufacturing* (Melbourne, Fla.: Krieger Press, 1973).
10. Alex Taylor III, "Why Toyota Keeps Getting Better," *Fortune*, November 19, 1990.

# CHAPTER 2
# INTRODUCTION TO BASIC PROCESSES AND BASIC TERMINOLOGY

The word *manufacturing* comes from two Latin words for hand and make. With the changing times and the changing style of manufacturing, the term may be a misnomer as fewer and fewer processes are labor-intensive. Surveys of American manufacturers show that in many cases less than 10 to 20 percent of the cost of goods sold is attributed to the cost of direct labor. In some radical examples, this number is as low as 1 to 5 percent. Manufacturing has changed from being the art of manual operations to a science of mechanized processes.

Manufacturing once had a sequential context, wherein the designer of the product would pass the design to the producer of the product. This was graphically portrayed by Nevins and Whitney in their text on *Concurrent Engineering*[1] and re-created with modification in Figure 2–1. However, to effectively advance these ideas and changes within the manufacturing concepts, we must be very clear in terms of the terminology and basic understanding of those processes.

## SECTION 2.1  INTRODUCTION TO BASIC PROCESSES

Manufacturing is a method of applying scientific techniques to convert raw material to a desired end product. This end product may be a gear, a shoe, or a watch. The key to successful manufacturing is to produce the part to desired specifications at the lowest cost in the shortest time possible that provides value to the customer. In Frederick Taylor's era, we made things and allowed others to buy them. The manufacturing world was successful because there was very little competition. However, in today's era, alternatives abound, and we must truly understand customer needs and then articulate products and services to meet those needs.

**FIGURE 2-1**
**The Traditional Product-Process Serial Development**

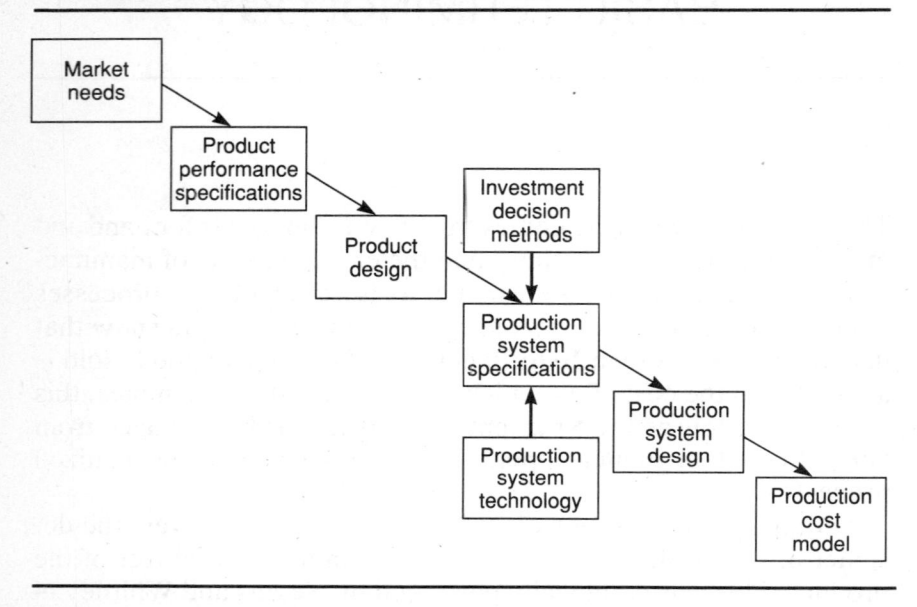

Source: Charles Stark Draper Smith Laboratory Results.

One methodology being espoused over the past few years is simply an attempt to compare what the product capabilities are to the customer requirements in a very formal manner. The methodology is called "quality function deployment" or QFD. A chart, similar to Figure 2-2, can outline and compare product features versus customer requirements. As these product features are compared, they become prioritized and gaps in product features are recognized. In a rigorous fashion, QFD can be applied to define how one can compare product features and capabilities to the customer's requirements.

## SECTION 2.2  BASIC TERMINOLOGY

Defining some key terms will help us communicate in a consistent fashion. The definitions are divided into three categories:

**FIGURE 2–2**
**A Simplified Quality Function Deployment Worksheet**

|  | **Features and functions of the product** <br> (The voice of the company) |
|---|---|
| **Perceived customer needs** <br><br> (The voice of the customer) |  |

1–5, basic terminology.
6–16, product design.
17–28, process design.

***Definition No. 1—Raw Material.*** *Raw material* is defined as purchased items or extracted materials that are converted via the manufacturing process into components and/or product.[2] The material is the critical first step in the processing of a product. The raw material is ordered from the supplier in a particular form. Often this raw material can be specified in terms of tolerances, material content, original shape, and so on. This is the first critical step in discussing the world of process design and development.

***Definition No. 2—Process.*** A *process* is a set of consecutive operations that completes a significant stage in the manufacture of components.[3] The process is the value-added function by which we add value to a component in a pattern pre-specified in a process plan. The adding of the value is the focus of this text.

***Definition No. 3—Tools.*** *Tools* are defined as those elements added to a machine tool or other work centers that adapts the machine tool or work center for the manufacture of a particular part or set of similar parts.[4] For example, jigs, fixtures, cutting tools, and numerical control tapes do not directly perform the process but indirectly are involved in the formation of a component.

***Definition No. 4—Equipment.*** *Equipment* is defined as all facilities provided to assist in the execution of manual operations.[5] These can be benches, hand tools, surface plates, and so on. The equipment is the support methodologies by which the facilities play a major role in the processing of the product.

***Definition No. 5—Work Center.*** A *work center* is defined as a specific production facility consisting of one or more people and/or machines that can be considered as one unit for the purpose of capacity requirements planning and detailed scheduling.[6] A work center is the fundamental building block on which the notions of processes and process planning are developed. The work center is the merger of a person and a machine in order to carry out the processes of today.

## SECTION 2.3   BASIC INFORMATION TERMINOLOGY NECESSARY FOR PRODUCT DESIGN

Critical elements of information are necessary to be clear on what the process plan is and how the process plan can be created. A few of these key definitions follow.

***Definition No. 6—Process Plan.*** The *process plan* is defined as the systematic determination of the methods by which a product is to be manufactured economically and competitively.[7] Usually, the task of process planning involves a series of steps. The first is active participation in the product design process to determine process feasibility and affordability. The next step is the interpretation of the design data including batch sizing, geometric configurations, raw material properties, dimensions and tolerances, and so on. What follows is the creation of the routing or "process plan" that is the recipe for producing the part. Specific details in terms of the specific tool paths, jigs, and fixtures follow until the part is produced.

**Definition No. 7—Production Plan.** A second fundamental information element in the information processes is the production plan. The *production plan* is defined as the agreed on plan that comes from the sales and operations planning function.[8] This is often used synonymously with *scheduling*. Production planning includes all aspects of management dealing with the various elements of the facilities themselves. It could be a corporation looking at the enterprise view dealing with the customer. It could be dealing with the supplier, or it could be focusing within the firm.

**Definition No. 8—Inventory Control.** A third information element fundamental to the success of process design is inventory control. *Inventory control* is defined as the activities and techniques of maintaining the stock of items at desired levels, whether they are raw materials, work in process, or finished products.[9] It could also be defined as the feedback control system that, by comparing actual stock levels with planned stock levels and feeding back information about excessive variances to the appropriate manager, makes it possible for the manager to take corrective action.[10] As we look at effective process management, the levels of inventory must be managed correctively. With too much inventory, we exceed our work-in-process cost variations; with too little inventory, the facility runs dry. The focus over the past few years has been to significantly reduce the inventory in the factory by migrating from a supply push environment to a demand pull environment. By being able to rapidly respond to changing customer demands, it is possible to develop strategies of production planning that allow us to make to order and respond very quickly to changing customer demands. This has a dramatic effect on inventory, as shown in this text.

**Definition No. 9—Method.** Method is another critical element in the definitions of information elements leading up to the process. A *method* is a planned sequence of work operations used to complete the manufacture of a part or product, or to complete a management task.[11] The method is critical because it is the formal documentation of what the sequence of work needs to be and the complexity and methods by which those tasks are performed.

***Definition No. 10—Bill of Material.*** Another crucial element of information is the bill of material. A *bill of material* is a listing of all the subassemblies, intermediates, parts, and raw materials that go into a parent assembly, and it shows the quantity of each required to make an assembly.[12] The bill of material is the listing of all parts in some sort of a structured, hierarchic manner that allows us not only to develop the individual process plans but also to amalgamate those individual processes and parts into an assembly.

***Definition No. 11—Master Bill of Material.*** Another element of this information context is the master bill of material. The *master bill of material* is a document listing the parts and the quantity of each part required to produce *one unit* of a particular product or subassembly.[13] This term is often used synonymously with the term *engineering bill of material*. It could also be compared to an order bill of material, where an order bill of material is a document listing the parts and the quantity of each part required to produce the number of units of a particular product or subassembly specified on a production order. As economic order quantities approach one, the master bill of material and the order bill of material become critical elements to compare and contrast as we look at the economics of world-class manufacturing operations.

***Definition No. 12—Plant Layout.*** Another element is the plant layout from an information standpoint. The *plant layout* is the physical layout, either existing or planned, of industrial facilities.[14]

***Definition No. 13—Process Chart.*** As we develop the notion of continuous improvement in the course of this text, process charts are fundamental to recognizing progress in the development of a process and how it can be improved. *Process charts* are those graphic representations of events occurring during a series of actions or operations, with information pertaining to those operations.[15] If we can develop a model to characterize each process, and if we can then develop a strategy for allowing individuals to recognize how to improve that process, we are employing the concept and notion of continuous improvement.

***Definition No. 14—Sequence.*** *Sequencing* is defined as determining the order in which a manufacturing facility is to process a number of different jobs in order to achieve certain objectives.[16] The sequence of products must take into consideration setups, changeover times, and more complex notions such as group technology, which is developed later in this text.

***Definition No. 15—Work Measurement.*** Finally, an element that is critical to success and not well understood in 21st century manufacturing is the term *work measurement*. Traditionally, work measurement has been conducted by industrial engineers who establish a time by which each process must be developed and also a learning curve by which the time for each element will be reduced as individuals learn how to perform that process. *Work measurement* is defined as the application of techniques designed to establish the time for a qualified worker to carry out a task at a defined rate of working.[17] To develop well-understood capacities and proper process designs, the issues of work measurement must be clearly specified.

***Definition No. 16—Economic Order Quantity.*** The final element in the information context of this text will be defined as the economic order quantity (EOQ). *EOQ* is defined as a type of fixed order quantity that determines the amount of an item to be purchased or manufactured at one time.[18] It could also be defined as the amount of an item to be purchased or produced at a given time to minimize costs associated with ordering and carrying expenses.[19] As this text evolves, the notion of an EOQ will take on new meaning as flexibility and efficiency are merged to allow an EOQ of 1 to be produced.

## SECTION 2.4  THE METRICS—DEFINITIONS FOR PROCESS DESIGN

Critical metrics exist to measure how effective we are at producing a product. Key elements used in this text are shown below.

**Definition No. 17—Downtime.** Downtime is defined as the time when a machine is scheduled for operation but is not producing for reasons such as maintenance, repair, or setup.[20] Planning for downtime is critical to the success and the efficiency of the equipment and the production process. Downtime also leads to a second key term—*idle time*.

**Definition No. 18—Idle Time.** *Idle time* is defined as time when operators or machines are not producing product because of setup, maintenance, lack of material, tooling.[21] In the *APICS Dictionary*, this is said to be synonymous with downtime. Idle time is critical for the measurement of the non-value-added time available for each product within the work force.

**Definition No. 19—Indirect Cost.** Another critical element of our metrics is indirect cost. *Indirect cost* is defined as those costs that are not directly incurred by a particular job or operation. Certain utility costs, such as plant heating, are often indirect.[22] Indirect cost has been growing at tremendous rates over the past 10 to 20 years. As we look at the cost of indirect manufacturing and overhead growth and as we compare it to the reduction of direct labor content, we find the ratio of indirect to direct costs is growing dramatically. This is reflected in our overhead rates. This also is reflected in the transition to new, modern accounting methodologies that can identify which of these indirect functions are truly adding value to the process.

**Definition No. 20—Indirect Material.** Another key metric is indirect material cost. *Indirect material cost* is defined as those materials that become part of the final product but in such small quantities that their cost is not applied directly to the product.[23] The indirect material costs added to the direct material costs gives the *total material costs* for the product.

**Definition No. 21—Throughput Time.** The next metric is the manufacturing throughput time. In this text, *throughput time* is defined as the average time taken to complete products within the factory—the equivalent to the number of days worked in a year divided by the physical rate of stock turnover.[24] In the *APICS Dictionary*, throughput time is synonymous with cycle time. But in recent industry advancements, the time from start to finish of a product has become of paramount importance. The *manufacturing throughput time* is defined in this text as the summation of all the processing times multiplied by a factor of $n$ where $n$ is the efficiency factor for that factory.

The equation reads as follows:

$$\text{Mfg. Throughput Time (within the factory)} = n \left( \sum_{i=1}^{m} (\text{Process times}). \right)$$

for all processes from 1 to $m$ on the critical path.

To measure the effective time a product is in the manufacturing state, it is necessary to measure the total throughput time. But much of this time is waiting for something to be done to the product rather than having value added. A representative metric for world-class companies is the efficiency factor $n$ that serves as a multiplier of the total of all the value-added process times to achieve the manufacturing throughput time. By world-class standards, $n$ is approaching 1.1 to 1.2, representing how efficient the company is at adding value. However, in many job shops throughout the world, $n$ may take on a factor of 10 or 100 or 1,000. As we target the value-added process times and the efficiency of those, the factor of $n$ becomes critical in the migration to world-class manufacturing.

**Definition No. 22—Product Throughput Time.** *Product throughput time* is defined as the total time taken by a product to pass through the enterprise between the first arrival of the first item of purchased materials used to make it and dispatch of the first finished product from the factory.[25] This is the first unit time to travel through the complete set of steps within the factory.

**Definition No. 23—Lead Time.** The *lead time* metric is the time span required to perform an activity.[26] Lead time is critical; although the processing times may be only a few days, the lead time may be a few weeks. That differential is causing significant concern as producers respond to changing customer demands.

***Definition No. 24—Work in Process.*** *Work in process* is defined as those products in various stages of completion throughout the factory, including raw material released for initial processing and completely processed material awaiting final inspection and acceptance as finished products.[27] Work in process is critically recognized as a metric that must be reduced in response to time as a fundamental metric.

***Definition No. 25—Capacity.*** *Capacity* is the aggregate number of units of work that can be completed at a work center in a given period of time.[28] Capacity is the absolute value of the value-added potential for each of the work centers. A goal is to optimize this value.

***Definition No. 26—Run Quantity.*** *Run quantity* is defined as the number of identical parts produced in a particular work center before changing to make some other part.[29] The run quantity can be as small as one or as large as millions.

***Definition No. 27—Setup Time.*** *Setup time* is the time required for a specific machine, line, or work center to convert from the production of one specific item to another.[30] As we begin to understand the notion of economic order quantities (EOQ) and watch these reduced to one, we find that the setup time factor for the economic order quantity must approach zero in order to achieve this goal. To achieve a setup time of zero, we have to employ new concepts, such as the notions of single minute exchange of dies (SMED) as referenced by Shigeo Shingo in his book.[31] This notion is displayed and developed later in Chapter 4.

***Definition No. 28—Tolerance.*** Lastly, *tolerance* is defined as the difference between the maximum and the minimum limits assigned to a dimension.[32] For example, we might be given a dimension of 25 inches plus or minus 0.2 inches. Because quality is truly defined by what the customer expectations are, the manufacturing world is migrating to different measures of tolerancing, as we will see later in this text.

## SECTION 2.5 INTRODUCTION TO TYPES OF MANUFACTURING

In Chapter 3, three different types of manufacturing are addressed. These three unique types of manufacturing are being developed throughout the United States and the rest of the world. We will describe how process design and development is used in a metal fabrication process and then extend the concept in very simple ways to nonmetallic and electronics manufacture. These three types of manufacturing will be called a metallic part manufacturing process, a nonmetallic part manufacturing process, and an electronic part manufacturing process.

For a metallic part, typically a raw metallic piece of stock is acquired, and material is removed to achieve the finished part shape. On nonmetallic parts, often called composite materials, the reverse process is frequently used, where individual laminates are laid one on top of the other in specific configurations to produce a product. Instead of removing material, material is built up through a series of layers.

In the area of electronics, typically the building block is a component, an integrated circuit chip, resistor, or capacitor, that is installed on a printed circuit board to produce a characteristic desired by the customer. The process is a combination of chemical processes used to produce layers on the semiconductor chip, followed by packaging and testing of the chip, followed by lamination and drilling of a printed circuit board, followed by the insertion of the components into specified geometric configurations that allow the activation of certain circuits within the personal computer board to elicit the correct response given a certain stimulus.

These three processes will be defined with a case study example in Chapter 3.

## REFERENCES AND RECOMMENDED READINGS

1. James L. Nevins and Daniel Whitney, *Concurrent Design of Products and Processes* (New York: McGraw-Hill, 1989).
2. *APICS Dictionary,* 6th ed. (Falls Church, Va.: APICS, Inc., 1987).

3. *APICS Dictionary of Production and Inventory Control Terms* (Falls Church, Va.: APICS, 1963).
4. Ibid.
5. John L. Burbidge, *IFIG Glossary of Terms Used in Production Control* (Amsterdam: Elsevier Science Publishers, 1987).
6. *APICS Dictionary,* 6th ed.
7. *APICS Dictionary,* 5th ed. (Falls Church, Va.: APICS, Inc., 1982).
8. *APICS Dictionary,* 6th ed.
9. Ibid.
10. Burbidge, *IFIG Glossary.*
11. Ibid.
12. *APICS Dictionary,* 6th ed.
13. *APICS Dictionary of Production.*
14. *Encyclopedia Dictionary of Production and Production Control* (Englewood Cliffs, N.J.: Prentice Hall, 1964).
15. *APICS Dictionary,* 6th ed.
16. Ibid.
17. British Standard 3138, "Glossary of Terms Used in Work Study and Operations and Maintenance," 1979.
18. *APICS Dictionary,* 6th ed.
19. David D. Bedworth and James Bailey, *Integrated Production Control Systems* (New York: John Wiley & Sons, 1987).
20. *APICS Dictionary,* 6th ed.
21. Ibid.
22. Ibid.
23. Ibid.
24. Burbidge, *IFIG Glossary.*
25. Ibid.
26. APICS Dictionary, 6th ed.
27. Ibid.
28. Ibid.
29. Burbidge, *IFIG Glossary.*
30. *APICS Dictionary,* 6th ed.
31. Shigeo Shingo, *A Revolution in Manufacturing: The SMED System* (Stamford, Conn.: Productivity Press, 1985).
32. Burbidge, *IFIG Glossary.*

# CHAPTER 3
# INTRODUCTION TO BASIC PROCESS DESIGN USING THREE EXAMPLES

Chapter 2 introduced the definitions that will be used in this book. An example will help understand the definitions. This chapter begins with a survey of "Lessons Learned in Implementing New Technology."[1] As is the case in Chapters 8 and 11, these surveys set the stage for the detailed case studies. This chapter also introduces three types of production to ground the reader on the principles and practices present in each. These three are metallic product development, nonmetallic product development, and electronic product development. We will develop the steps needed to define and develop these in an integrated environment.

The survey was published in 1989 in an attempt to assess what needed to be done for a successful technology installation. The survey consisted of the following respondents:

- John Deere Company.
- FMC Corporation.
- McCord Gasket.
- Eastman Kodak.
- Georgia Tech.
- Remmele Engineering.
- J. P. Industries.
- A. O. Smith.

Five key elements were consistently found throughout the successful installations. These five are:

1. Educate! Educate!—everyone must be clear of the vision, the missions, the plans, and the actions as well as well-versed in the tools needed for success.
2. Think organizational change—any infusion of technology or change of business practice will force change.
3. Match the technology to the needs—a simple thought. Don't overkill or underkill. It is mandatory for this match to meet the needs yet be culturally and intellectually manageable.

4. Instill confidence—as the project team is launched, management must demonstrate trust and confidence.
5. Don't force technology—there is a cultural acceptance lag between the state of the art and the company. An alignment of the position of the company with the technology availability was found to be critical.

This chapter shall describe how a simple technological infusion at Rockwell International led to some startling results. But the five points above must be emphasized. At Rockwell, simplicity and standardization were culturally acceptable—not whizbang. The case exhibits these qualities.

## SECTION 3.1  AN EXAMPLE OF A METALLIC PART PROCESS DESIGN AND DEVELOPMENT

Process design for metallic components dates back hundreds of years. In the early years, basic metals were used to forge or cast a frame or somehow develop a product from a piece of metal. This concept was fundamental to the success of the Industrial Revolution. The product was normally designed first and then, through a series of stages called conceptual, preliminary, and detailed design, the design was formulated, prepared, and presented to a manufacturing engineer who would take the product definition and develop the process design. The process design would interpret the product definition and translate that product into a series of processes that would be performed on a piece of equipment or series of pieces of equipment to achieve the form, fit, and function required by the product definition.

But in today's environment, as developed later in this text, this serial activity is transformed into a parallel or concurrent activity where product definition and process definition simultaneously are fundamental to success, and time to market is of critical importance. For this to happen, it is necessary to look at the form, fit, and function required for a particular metallic product as it goes through the design stages and then translate those into the required manufacturing processes.

The case study herein was developed at Rockwell International in Dallas at the Collins Transmission Systems Division in the early 1980s by a team of cross-functional experts led by Mr. Arnold Kriegler, director of production operations.[2] The situation is unique to American industry. At the time of the case study, Rockwell had a significant backlog for the microwave radio system necessary for long-line communications. The management team believed cycle-time reduction was critical for the success of this product line over the next few years. To address cycle time, a model of the cycle times for the entire process was created and key opportunities were identified. The design-to-manufacturing cycle time was crucial, but once designed there were key components that formed the critical path. Because every radio was to a unique specification, certain key parts could not be made before the design was completed. One of these crucial parts was a waveguide. This unique part was the glue in the microwave radio system because it was used to guide the electronic communication waves between various elements of the radio in its 19-inch rack. The part is simply a rectangular-shaped tube of a special metallic component that had soldered to both ends a curved section that allowed the waveguide's flanged ends to connect to the radio. On analysis of the waveguide situation, the following facts were found as shown in Figure 3-1.

The fabrication shop at Rockwell probably represents every fabrication shop in the nation in that small orders were often launched with inadequate lead times and the name of the game was "expedite!" But the paradigm of management under the direction of Mr. Kriegler began to switch when "time" was identified as one of the crucial yardsticks for success. Once the times of the processes were modeled and the layouts were analyzed, the cross-functional management team found the results shown in Figure 3-2.

To attack the "as is" situation at Rockwell, the team looked at the overall product-process life cycle and determined that several changes must be made—and not all were in manufacturing. The first step was to ask the product design team some very fundamental questions. One of the most revealing answers came to the question, "Why are the tolerances to three decimal place accuracy?" The answer: "Because we always have!" Analysis of the manufacturing process revealed the first operation is a sawing operation, and this was the primary cause of poor dimensional quality of the product.

**FIGURE 3-1
Fabrication Shop Facts**

- Operates as an internal supplier of mechanical piece parts and assemblies to manufacturing.
- Metal forming/cutting job shop, with machines/processes functionally grouped within 70 work centers.
- 175 major machine tools, occupying 100,000 square feet.
- 400 employees, 325 direct production labor.
- 4,000 open work orders at any given time.
- 14,500 work orders completed per year on 12,500 different part numbers.
- Average order size less than 50.
- 55% of orders are for quantities of less than 10.
- 50% of orders are initiated with inadequate lead time.

When the cross-functional team asked the design engineering team if the length of each of the waveguides could be standardized to 1/10th-inch increments on the cutting fixtures needed to achieve the tight tolerances, the design team said, "Sure!" *Product-process issue No. 1 was resolved.*

Next, the team looked at the supplier base because it was absolutely necessary to have raw stock in the correct configuration at the correct inventory levels for cycle times to be reduced. The cross-functional team next went to purchasing, where supplier selection of waveguide raw stock was discussed. Purchasing staff members said they had found one "great" supplier of waveguide stock and were willing to take a chance to structure a strategic alliance with this supplier and make it the primary source of waveguide stock. *Product-process issue No. 2 was resolved.*

Finally came the analysis of the processes performed to create the piece of waveguide in the fabrication shop. As we learned earlier, the fab shop was a typical arrangement of diverse pieces of equipment organized by their function. In developing the "as is" process flow shown in Figure 3-2, we learned that it took 23 moves and 17 weeks to produce the product.

But now the detailed analysis of the process began. Through

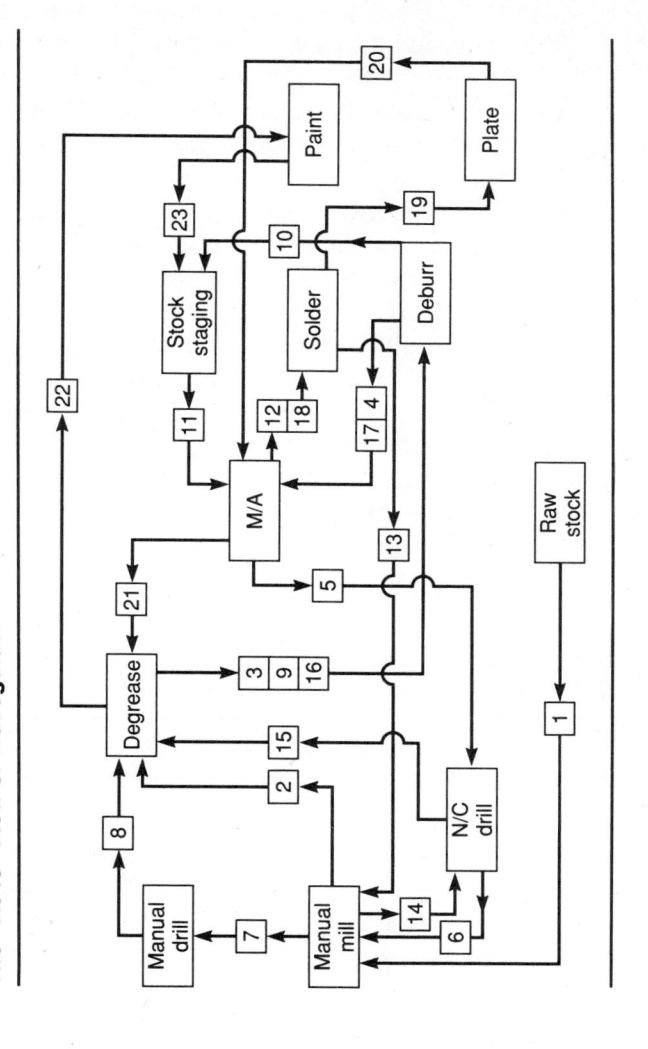

**FIGURE 3–2**
**The "As Is" Flow of Waveguide**

*simplification,* common tooling and fixturing could be developed for this "family" of products that allowed smooth flow. Through *consolidation,* the waveguide processes were brought together into a defined area of the factory that became the waveguide cell. Through *elimination,* certain steps of the process routing were changed or eliminated to achieve better flow. The final layout is shown in Figure 3–3. Note that now the flow contained only nine moves and could be achieved in 1.7 weeks of throughput time. *Product-process issue No. 3 was resolved.*

Three startling results occurred after the cell was installed and the integrated product-process design and development were completed. The first, the financial results, is shown in Figure 3–4. Through audited accounting of the total process design and development, the waveguide cost was reduced by 28 percent and the payback was measured in months. But two even more dramatic events actually sealed the success.

**FIGURE 3–3**
**The "To Be" Flow of Waveguide**

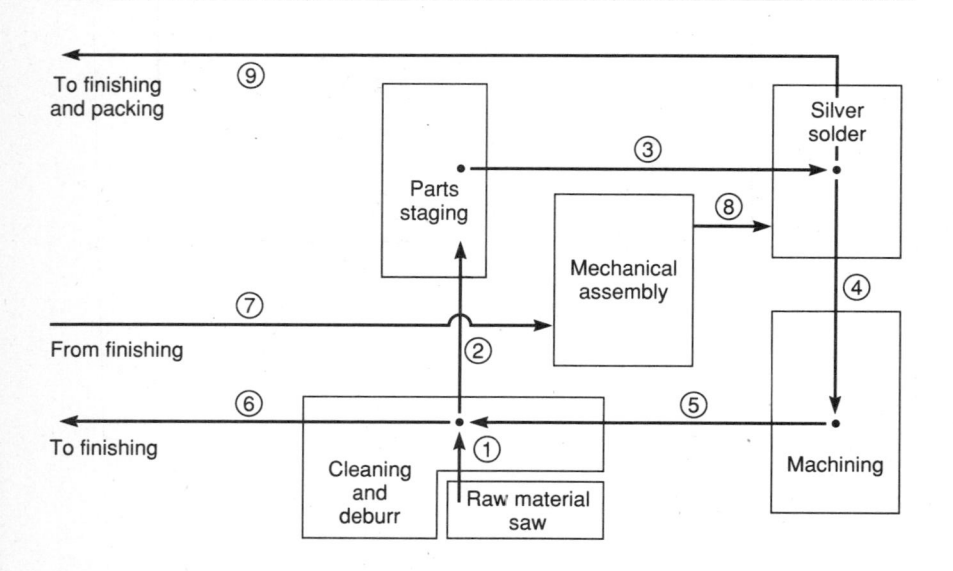

**FIGURE 3-4**
**Financial Results from the Waveguide Cell**

| Improved | + | Reduced | = | $avings | 28% |
|---|---|---|---|---|---|
| Quality ▲ 75% | | Setup time ▼ 10% | | Part costs | |
| Performance ▲ 17% | | Run time ▼ 10% | | WIP inventory | |
| | | Inspect time ▼ 60% | | Material inventory | |
| | | Throughput time ▼ 71% | | | |

The first dealt with the impact of the cell on the material management team. Before the cell, the team had to forecast the need for waveguide because of the 17-week lead time. This did not allow sufficient time between when the customer placed the order and when assembly needed the fabricated parts. But with the cell and the standardization, a 1.7-week throughput time allowed a material management change of strategy from "make to stock" to "make to order," and small economic order quantities were now placed in orders only in a "firm" condition.

Finally, the true test of the cell was the morale of the team operating the cell. In studies conducted by independent organizational development experts, the morale of the operations team members was higher than their peers when the cell was created and remained that way for at least five years.

The learning model used here is the same in the 1990s as it was in the 1980s:

- *Simplify* with consolidation and elimination.
- Then *integrate* now knowing how best to perform each task.
- And, finally, *automate* if it makes sense.

The analysis phase begins with simplification through the creation of a model of the product-process environment. As detailed in Chapter 4, it is first necessary to define and understand how the process needs must conform to the product requirements in terms of form, fit, and function as they relate to making the customer successful. Chapter 5 develops the overall design of the process. For our particular sample part here, the design of the process was developed in a concurrent engineering mode in the overall product-process development cycle. As we learned, the notion of concurrent engineering is fundamental to the success of time-to-market compression, and several examples are given of how time to market can be condensed using modern notions. We will also find that although the cost appears to be much higher when we slow the early stages of design in order to accommodate process as well as product, the payback is often remarkable.

The notion of group technology (GT) is developed in Chapter 4 as we begin to identify simplification, elimination, and consolidation opportunities by attempting to standardize and take advantage of similarities of process needs and process requirements as they relate to the overall form, fit, and function of this particular part. The notion of group technology is very simple: For a product definition or for a process definition are there similar parts in the part library that can serve as a baseline by which we can simply modify an existing process plan as compared to developing a new process plan? Benefits of this and the foundation on which the focused factory can be developed are also developed in Chapter 4.

The last analysis step in simplification is to simplify tooling. We wish to run in a just-in-time environment for the overall design of the product. To achieve Kanban, or just-in-time philosophies, the setup time of tooling and fixturing of the process for this particular product must be very, very small. Shingo Shigeo developed in his text several years ago the notion of a "single minute" exchange of dies. The tooling and the fixturing for this particular part is nontrivial. Simplifying the tooling and fixturing must be emphasized to achieve standardization and simplification that allows an economic order quantity of one and requires as a prerequisite that the setup times *must* approach zero. To do that, one must emphasize the notion of simplification first, followed by integration, and ultimately automation. Chapter 4 is dedicated to understanding these notions.

Chapter 6 outlines how this part family can be developed. First, the overall process plan for the part is developed, then the geometries for the part are defined. Then something called the cutter location data file for all the processes that are to be numerically controlled is defined. We will also then develop the computer numerical control methodology and how closed-loop control can be developed for this particular part.

Finally, Chapter 7 looks at the metrics necessary to achieve this. We find that in normal circumstances, lengths, diameters, speeds and feeds, and depths of cut for the particular processes are examined. But more importantly, we must look at traditional metrics of cost and schedule and quality as well as some new metrics for today, which say, in essence, that if *time* is the new metric, then we must focus on time and all aspects of time. In his classic performance assessments in the 1960s and '70s, Dr. Eugene Merchant pointed out that fully 95 percent of the time that a work piece is on the factory floor nothing is being done to it. It is sitting idle, and there is no value-added process. And of the 5 percent that remains, approximately 1.5 percent of that time it has value being added and 3.5 percent of the time it is in motion or is being set up or torn down. The new metrics of today will focus on market share, time-to-market compression, the notions of total quality, and how inventory management and the implications of process design must be understood. Finally, Chapter 7 discusses historical usage of the learning curve and the modern transitions to the usage of the learning curve.

## SECTION 3.2   AN EXAMPLE OF A HOW A NONMETALLIC PART IS PRODUCED

Contrary to the approach taken in Section 3.1, there are differences in how a nonmetallic or a composite material part is produced as compared to a metallic part. However, the same early steps are required to define the form, fit, and function, and the various prototyping steps are necessary.

Composite materials possess certain properties that make them attractive alternatives to metallic structures. The most important is the strength-to-weight ratio of composites in certain dimensions. The composite gets the name because it is made of multiple materials, often in a matrix structure that begins like a piece of tape or a sheet of cloth. When tape-like rolls of composite materials are used, a machine unrolls the ribbons in fixed patterns to form a laminated sheet in the size desired. By mixing the orientations of these plies, certain characteristics can be predicted. When broad goods are used, a precision cutting system is employed similar to the way a

**FIGURE 3–5**
**Flexible Composites Center**

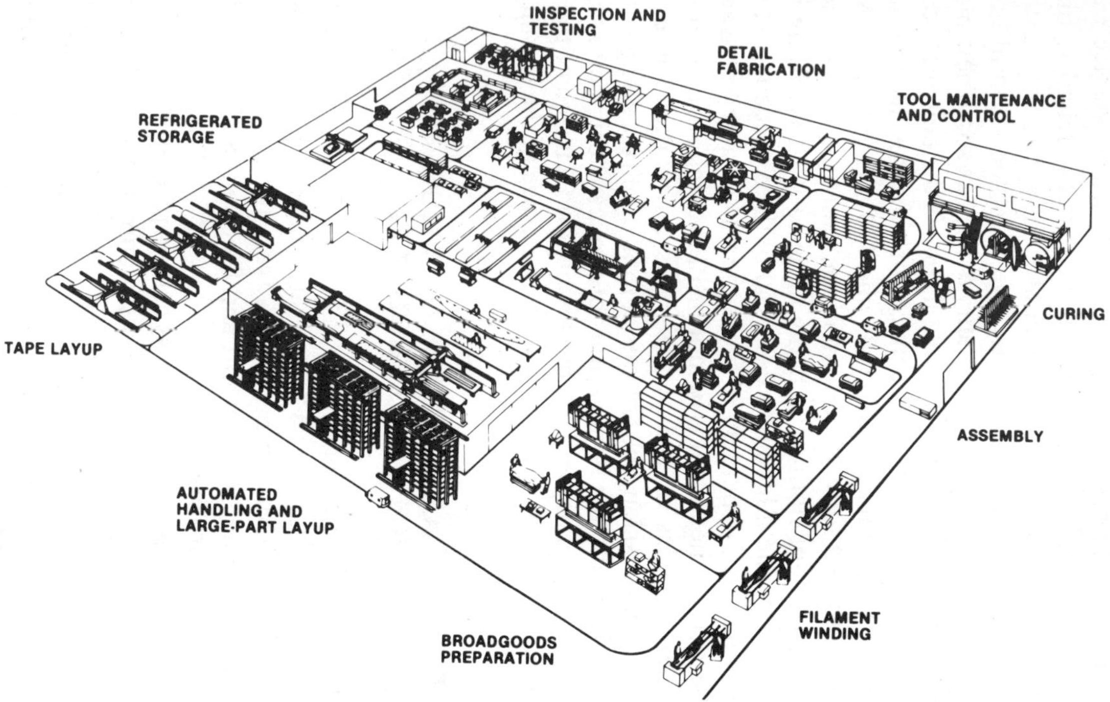

tailor would cut a suit from a roll of fabric. And just as the tailor is concerned about having the patterns match when the sleeve is connected to the body, the composites "tailor" has the same concerns.

Also unique to the composites material is its need to be chilled before curing to preserve its lifetime. Figure 3–5 shows a composites center. This comes from an Air Force-funded effort at Vought Aerospace in Texas.[3]

But in composites, just as in metallics, the same principles of product-process integration apply. The earlier the process team can participate in the design, the better the product and the more producible the product. Geometry is still a concern because now the plies must align, and dimensioning and tolerancing continue to be a challenge. But the added complexity of shelf life presents a new scheduling issue for the planners of nonmetallic production. Shelf life is the time allowed for nonmetallic materials to be processed. Unlike metallic products that can sit in inventory for months, many nonmetallic materials have shelf lives measured in days (e.g., 1,000 hours) with the time available to work on the materials outside a refrigerated environment measured in minutes or hours.

## SECTION 3.3 AN EXAMPLE OF HOW AN ELECTRONIC PART IS PRODUCED

The combination of metal removal processes, such as drilling and milling, and composite lay-up processes, which align one layer of material on top of the next, creates complexity in electronic products. The line widths of the semiconductor lithography is now approaching less than 1 micron in width, yet in some semiconductor chips, 30 or more layers of material must be stacked within very tight alignment dimensions. But just as with metallic and composite materials, the principles apply. Because there is such a high degree of advanced process technology required to produce a semiconductor, this may be the supreme challenge for product-process design and development integration.

## REFERENCES AND RECOMMENDED READINGS

1. Robert Stauffer, "Lessons Learned in Implementing New Technology," *Manufacturing Engineer,* June 1989.
2. Dan L. Shunk, "Group Technology Provides Means to Realizing CIMS Benefits," *Industrial Engineering,* April 1985.
3. A. J. Roch, Jr., "Productivity by Design-FMS Applications That Work," Dearborn, Mich.: FMS Conference, Society of Manufacturing Engineers, 1986.

# CHAPTER 4
# PREREQUISITES FOR INTEGRATED PROCESS DESIGN AND DEVELOPMENT

Before discussing the various aspects of integrated process design and development, certain fundamental prerequisites must be in place to understand the directions in which technology is progressing. The first notion is that of concurrent engineering and how teams of talented individuals are forming product and process action teams. But in many cases today, the teams are coming together virtually from sites around the world. An example of this is the way Digital Equipment Corporation developed its award-winning RA90 storage module. This module had design teams from the Near East, Europe, North America, and the Far East all working on the design in a virtual way. Although this may not be representative of all products today, globalization of products will force this trend in the 1990s.

The second is the concept of group technology where taking advantage of similarities becomes a science, not an art, and where we can identify similarities and be able to capitalize on those from a variety of viewpoints. Third, the notion of total quality management (TQM) is viewed as the continuous improvement foundation based on Deming's work and captured in a Department of Defense program plan.[1] Fourth is the notion of material management and responsiveness from a time standpoint in a just-in-time strategy using Kanban scenarios. Fifth is process capability design including supplier benchmarking and material benchmarking. Sixth is the integration of material handling and material flow systems into the overall integrated process. Here, notions such as simplification, integration, and automation are introduced as a fundamental approach to systems work. These are reinforced in Section 4.7 where simplification of setup processes can dramatically change the overall capability of the factory. Next is the introduction to systems integration; Section 4.9 introduces a methodology to model and design integrated process systems from a total, global viewpoint. Finally, we view the tool called simulation and determine how it can be used. These and other aspects will be discussed in this chapter in anticipation of developing the integrated process design framework.

## SECTION 4.1   CONCURRENT ENGINEERING

Fundamental concepts of concurrent engineering are coming to light in the 1990s. Several notions such as the ability to capture and understand what the customer actually wants is one fundamental building block. The second is the ability to have many people come together representing a variety of functional elements and enterprise viewpoints to discuss the design of the product and the process in a simultaneous fashion. The third are methodologies for designing for producibility and also designing for assembly.

The first notion deals with how the conventional product design/process design and development scenario is changing into a concurrent environment. As Figure 4–1 illustrates, Dr. James Nevins and Dr. Dan Whitney from the Charles Stark Draper Laboratories have shown that the sequential series of events leading from market needs through product performance specifications to the design of the product and then ultimately the production system design have in the past been a serial activity, where one step precedes the other in a stepwise fashion, resulting in significant time delays and costs that ultimately translate into lost market share in today's environment.[2]

They contend this serial connotation must be transformed into a roundtable type of scenario, represented graphically in Figure 4–2, where people representing the multitude of functions from the enterprise come together early in the design stage to review the product requirements, compare those with the customer requirements, and ultimately develop the process design for the product.

To do this correctly, one must begin with a true understanding of what the customer wants. A structured approach must be in place to capture real customer needs. Evolving in modern literature today is a methodology called quality function deployment (QFD). This technique is simply a matrixing technique that compares what the perceived customer needs are against the features and functions of the product that the enterprise has in mind. As we develop this matrix, we use multiple inputs to address what the true issues are

**FIGURE 4–1**
**The Traditional Serial Approach to Product and Process Development**

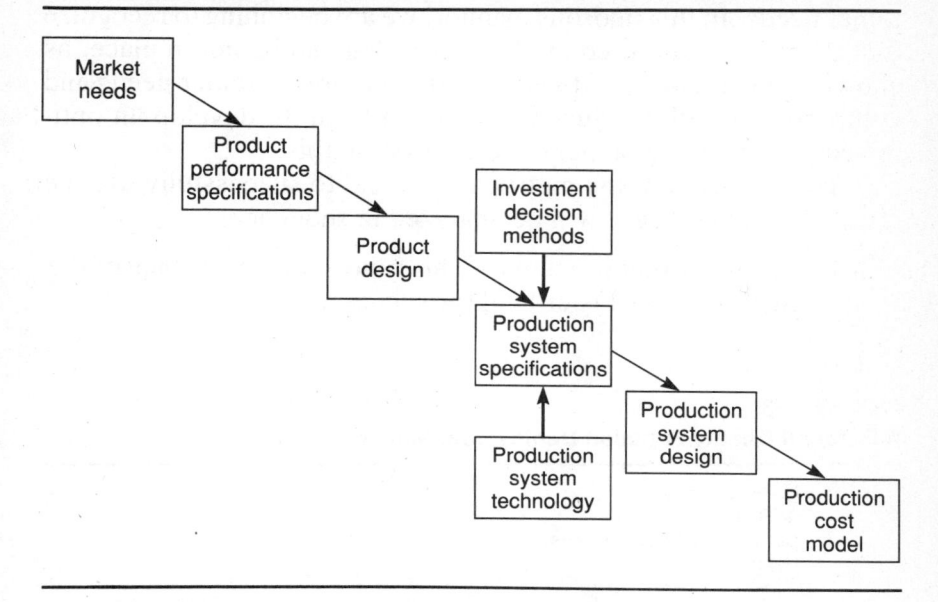

Source: Charles Stark Draper Laboratories results.

**FIGURE 4–2**
**The Concurrent Engineering Roundtable**

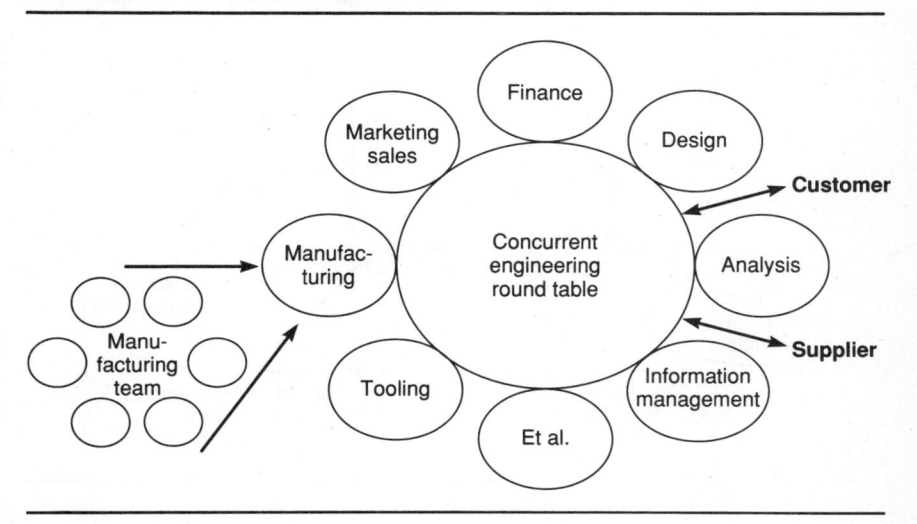

and then we begin to prioritize the product features per the customer demands and compare those features with the overall customer needs. In this rigorous manner, we are beginning to recognize that there is a structured methodology that can be put in place, as shown in Figure 4–3, that allows us to capture customer needs and compare that with product features and begin to develop an optimized product development cycle based on this.[3]

The elements of the matrix are critical to successfully use the QFD technique. These will be analyzed in sequence:

1. Customer requirements—This represents the "voice of the customer" and is critical for success.

**FIGURE 4–3**
**A Detailed Quality Function Deployment Matrix**

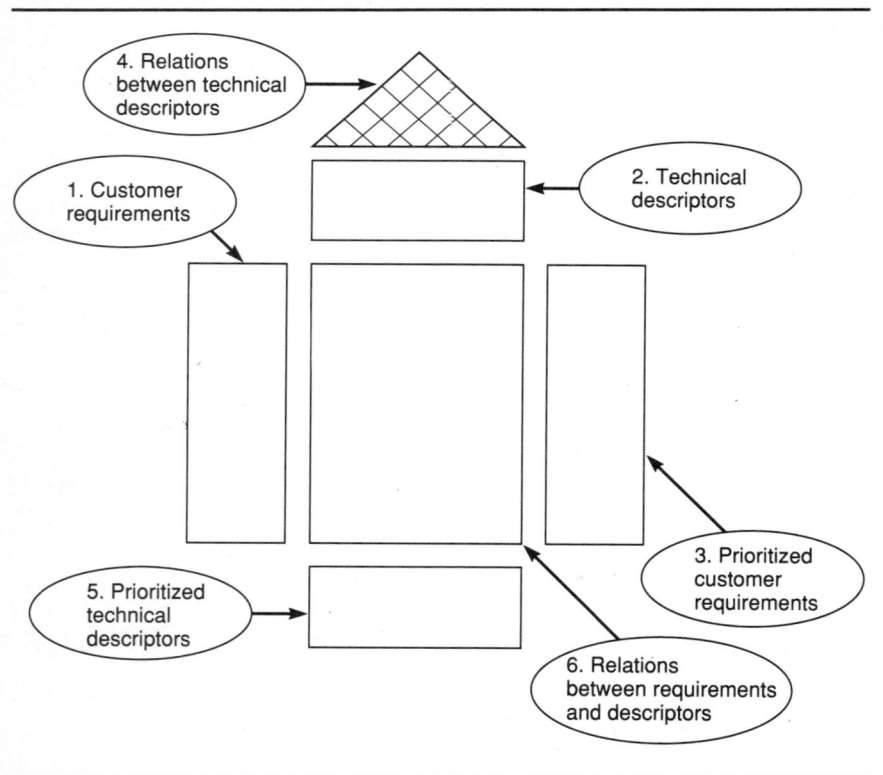

1. Customer requirements
2. Technical descriptors
3. Prioritized customer requirements
4. Relations between technical descriptors
5. Prioritized technical descriptors
6. Relations between requirements and descriptors

2. Technical descriptors—This represents the "voice of the company" and attempts to capture the features that the product brings to the customer.
3. Prioritized customer requirements—Somehow a prioritization must be achieved to assess which technical descriptors best meet the customer's needs and which are extras. The ratings of this can be negative or positive.
4. Relationship between technical descriptors—This internal assessment is conducted to ascertain any dependencies between the functionalities that have been defined.
5. Prioritized technical descriptors—Based on a product strategy, the technical descriptors must have some ranking of perceived value. These may be for differentiation purposes, low-cost purposes, or niche market purposes. Each must be assessed.
6. The relationship between requirements and descriptors— This is the critical heart of the QFD matrix and the heart of the exercise. Based on a weighted scale of any type, the customer view must be compared to the company view. This *must* be a realistic assessment if the product is to be successful.

The foremost test in quality function deployment is the ability to determine what the customers truly want. Any method that can lead to true customer needs definition must be employed. Interviews are a good methodology. Focus groups are a second. Surveys, observations of similar products, even inviting the customer to participate in the design of the product are paying huge dividends for companies. The bottom line for quality function deployment is that we must pay very strict attention to the *process* of design as well as the end results of the design. The notion of continuous improvement must be recognized throughout all of the process improvement steps. This notion is developed in Section 7.3, but a brief insight is in order here. Continuous improvement is dedicated to making small, incremental advancements in the product and process to create a sense of continuous change. By fostering continuous improvement, General Electric in its Appliance Park center in Louisville, Kentucky, has achieved a 5 percent productivity improvement each year for the past several years in a consumer product that is quite mature. The focus on compounded continuous improvement has been the hallmark of the Japanese thrusts in manufacturing.

The second notion of concurrent engineering is that of developing a closed-loop feed-forward feedback system as shown in Figure 4–4, which allows various members of functional elements to sit around a design roundtable at the early conceptual and preliminary design aspects of the product and recognize and translate true customer needs and wants and create in an interactive session a robust product definition and a robust process design in almost a simultaneous fashion. Concurrent engineering hinges on the notion that people of various disciplines can have the ability to interact with one another through a live environment, possibly throughout the world, to develop a much better product definition.

The third notion is the ability to design for what are called the "ilities." Here, Design for $X$, or DF$x$, where $x$ could be assemblability, viability, controllability, producibility, serviceability, survivability, any "ility" from a to z, is the essence of our objective. *Business Week* showed some significant examples of how this design for the "ilities" in concurrent engineering can provide sig-

**FIGURE 4–4**
**The Closed-Loop Feed-Forward/Feedback Concurrent Engineering Environment**

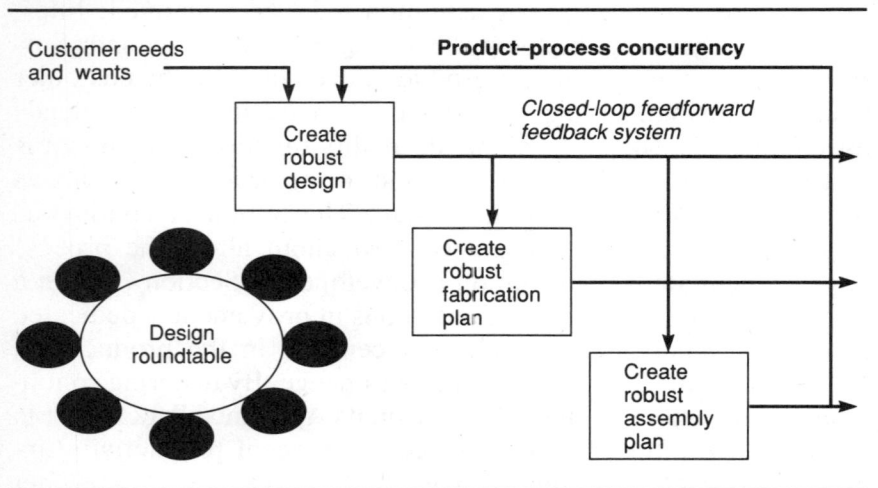

nificant benefits to the corporation.[4] For example, NCR, which produces terminals for checkout counters, started a concurrent engineering effort in 1987. Now a new product comes out in 22 months, as compared to 44 months in the previous methodology, with 85 percent fewer parts. It is now assembled in two minutes, which is a 75 percent reduction over the previous methodology. AT&T, using a concurrent engineering approach, has developed telephone switching computers, trimming the cycle time by more than 50 percent and plunging manufacturing defects by 87 percent. Finally, John Deere reduced a seven-year cycle time for product development by 60 percent and overall development costs by 30 percent for its agricultural and forestry products.

These results show that startling, significant achievements can be made in applying concurrent engineering to the overall product-process development cycle. The integrated process design and development team must understand the notions of concurrent engineering and how these notions can be brought to bear. One must also recognize, as we've discussed previously, that "hurry hurry hurry" is not necessarily the modus operandi for achieving this. We will display this in Chapter 7 as well.

The final issue that we need to review as we discuss concurrent engineering is the implications on the people. Team formation and team operation are nontrivial and, as Figure 4–5 points out, there are significant management responsibilities early in the creation of a

**FIGURE 4–5**
**The Issues in Migrating to Teamwork**

| Organizational Aspect | Traditional Practice | Shift in Practice |
|---|---|---|
| Authority | Based on position | Based on knowledge |
| Decision making | Close to top | Where action is |
| Employee contributions | Limit knowledge and skills | Enhance |
| Information | Closely control | Share widely |
| Rewards | Individual performance | Teamwork |
| Status | Highlight differences | Mute differences |
| Supervision | Watchdog | Resource |

Source: MSB Report of National Research Council.

concurrent engineering team that cannot be overlooked. Issues, such as the strategy for team operation, the boundaries of the team, the metrics of the team, the overall working philosophy of the team, and the chartering of the team must be specified by a management team, and then the authority and the responsibility must be handed over to the concurrent engineering team.

As the concurrent engineering team begins to take shape, management then must become a cheerleader and provide the resources necessary to achieve a success that has been dictated and defined. Team metrics must be very clear. How the reward structure for the team occurs is also being reviewed by many management teams throughout the country. The reward structure for concurrent engineering teams must be based not only on financial but also on nonfinancial rewards, and as Ann Majchrzak points out in the Computer and Automated Systems Association's Blue Book on teamwork,[5] these rewards must be for the individual as well as the team and must be consistent with the overall enterprise desires.

But the final major hurdle to concurrent engineering is the migration of management into a mode of distribution of authority while maintaining responsibility. This is the antithesis of how management has operated in the past and this fundamental shift is forcing significant changes in the management structures of many companies. As Charles Savage points out in his text on *Fifth Generation Management*,[6] the human networking that concurrent engineering is causing will create a shift in corporate culture of every company that deals with it.

The conclusions from concurrent engineering are that one must truly understand the real customer desires and needs; this is critical for the success of the overall process design. Time compression is truly an important metric that must be understood and may not be achieved the first time concurrent engineering is undertaken. But as we learn and improve, the second, third, and fourth opportunites for this should be much more successful. Cross-functional, virtual task teams are critical to the success of concurrent engineering. The creation of the prototyping, as will be discussed in Chapter 5, can take on many forms, and we must understand what types of prototypes will be required in a concurrent engineering mode.

In terms of design for assemblability, Boothroyd and Dewhurst,[7] Bart Huthwaite,[8] and others have espoused that the best part that exists in an assembly is one that is not there. The process of looking ahead in the design phase of a product to assess the ability to assemble that product becomes absolutely crucial for the overall success of the concurrent engineering process.

Finally, we must understand the impact concurrent engineering is going to have on people. Bringing together diverse groups of talented individuals and assigning them to a team is difficult. The first issue faced is simply the dynamics of the team. Most of the participants have been measured previously by individual metrics that forced the individual to optimize on his or her activities. With a team structure, the team metrics force synergy and compromise, and no one individual is able to dominate the sessions—at least in theory. Forming a successful concurrent engineering team requires the same patience and support that forming a successful basketball team must have. All of the team members must have a clear, consistent vision of how team success can lead to their success. The rewards must be a balance. Individual creativity is an asset of all corporations and must be nurtured. But this also must be tempered by the symbiosis needed for a successful team. The impact on people is tremendous. This is summarized in the trends for concurrent engineering covered in Section 10.1.

## SECTION 4.2   THE CONCEPT OF GROUP TECHNOLOGY

Group technology is employed every day in every enterprise. It is the simple notion that there is an operating management philosophy in place that recognizes that similarities occur in the design and manufacture of discrete parts. The identification of similarities is a fundamental building block to the migration of companies that standardize and simplify ways of doing business. It is also the foundation of focused factories, mentioned many years ago by Dr. Wickham Skinner[9] in his classic article on the migration of focused factories and the basis for cellular manufacturing for material flow and other aspects.

The notion of group technology says similarities occur in the design and manufacture of discrete parts. However, similarities can be construed in a variety of manners. For example, as shown in Figure 4–6a, Alex Houtzeel and C. S. Brown[10] wrote that design

**FIGURE 4–6a
Similarities Based on Geometry**

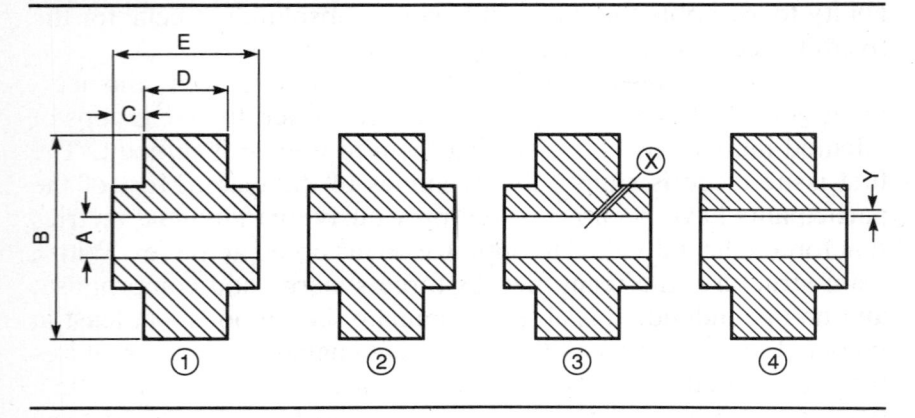

could be construed as having similar geometry. Whereas in a manufacturing sense, as shown in Figure 4–6b, Houtzeel and Brown recommend that similarities not be based on geometry, but rather on manufacturing processes should be used.

Furthermore, if we view the overall enterprise and if the cost of goods sold reflects a 50 or 60 percent cost for material acquisition, then we may have a third or a fourth view of similarities based on the overall supplier base and how raw material stock is acquired in the facility.

In any case, our first objective for group technology is to establish a structured, standardized way of acquiring and processing materials to achieve significant "economies of scale" even though we produce engineered products in an "economies of scope" environment.

Second, we wish to take advantage of similarities to be able to find parts that have been invented and designed as compared to designing new parts. Third, we wish to develop consistent, single-, and best-process plans for each of the parts that need to be produced. Fourth, we wish to regulate the purchase of components with total visibility based on economics, timeliness, and so on in a consistent manner. The bottom line for group technology is we wish to achieve flexibility in the face of change that will be ever present. With group technology, we look at the various functions throughout

**FIGURE 4–6b**
**Similarities Based on Process Requirements**

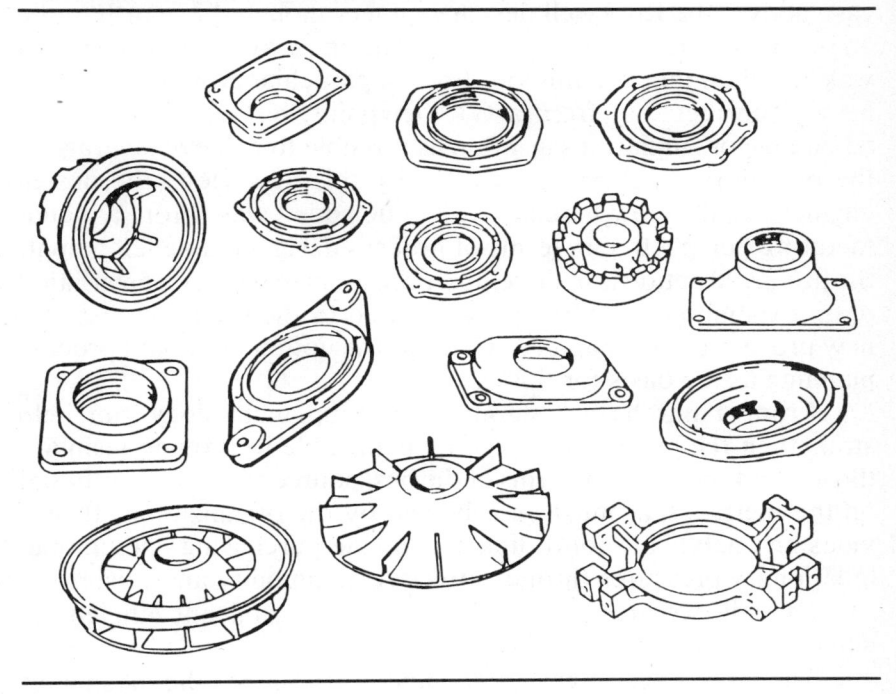

the enterprise and attempt to identify where similarities can be achieved.

For example, in the design function, we recognize that based on geometry, functionality, footprint, and/or a variety of other attributes about a product, we can take advantage of similarity by finding parts already in the design library. The economics of this are startling. A survey conducted by Rockwell International in the 1980s found that it cost the company $5,000 to $10,000 for each new part that was designed and put into the design library. Assuming your company produces 1,000 parts a year and of those, 5 to 10 percent are absolute duplications and another 10 to 20 percent are similar parts, significant economies can be accrued even for small businesses, which seldom recognize this in their existing environments.

Group technology forces the analysis team into recognizing

what is truly a value-added activity and then presents a tool that allows the team to optimize on these value-added functions. In the case above, the Rockwell design engineer thought his worth to the company was to design new parts. And the way he was rewarded was based on this assumption. From a parochial design viewpoint, he was correct, but from a GT viewpoint, his task was to take advantage of similarities as often as possible to achieve leverage in the overall product-process cycle. By this new view, the design engineer could see where his real value added was. From a manufacturing perspective in terms of process design and development, the preferred action is to develop a process plan based on a variation of an existing process plan as compared to the total creation of a new process plan. Chapter 5 develops the notion of variant process planning as the basis for this.

From a purchasing standpoint, group technology can help strengthen the buyer's leverage by being able to provide visibility throughout the world in terms of the best sources for products based on the metrics that must be achieved by the organization. It provides an analytical tool to link worldwide purchasing and combat inflationary pressures through effective commodity analysis.

Overall, we are attempting to achieve significant efficiencies while maintaining significant flexibilities.

The mechanics of group technology, similar to the mechanics of concurrent engineering, are becoming well understood in the research community. One must be able to identify a part family based on a multitude of approaches. It could be the geometry; it could be the material; it could be the form, fit, or function of that particular product. Four methodologies are present today to look at and identify where similarities can occur. The first is simply called visual inspection, or "occular group technology." In occular group technology, we simply walk around and identify where parts look to be similar. It is not rigidly scientific, and it is difficult when there are many parts. But it is a starting point by which one can begin to judge whether or not group technology seems to make sense in your environment.

The second, as espoused by Burbidge and others, recognizes that if existing process plans do exist within your corporation, you can begin to identify where similarities of production flows occur using the methodology Burbidge calls "production flow analysis."[11] Burbidge simply compares the process routing for each part fabricated in a particular facility. If the process routings are similar, then the production flow analysis that Burbidge espouses dictates they must be similar parts from a process viewpoint and they could be linked to form a manufacturing cell or a focused factory.

The third approach for the mechanisms of group technology has a bit more rigor to it. It's called "classification and coding." It begins by identifying some key intrinsic qualities or characteristics about a product called "attributes"; these could be geometry, processes, or form, fit, or function. These attributes are then *classified* in terms of groupings of similar attributes of components.

Then, a symbol or a *code* is given to each of these classification systems. For example, the Dewey decimal system was created to find a book in a library. If we have a classification, let's say "Management of Production," and a code for that, 658.5, we could easily find a book on this subject in any library. The same is true for ZIP codes in the United States. Given 46391, we know we are probably in northern Indiana and we can zero in on the city of Westville by simply knowing 46391 and the ZIP code context. Many approaches have been developed for storing codes based on whether they should be related one digit to another or independent of digits. Coding systems that relate one digit to the previous one are called "monocodes," while those coding systems that allow independence of digits are called "polycodes." Most corporate classification systems are hybrids with mixtures of monocode and polycode segments.

Manufacturing companies often create a "commodity code" either in the part number or attached to the part number that allows the company to identify similarities. This is an excellent start with data and a data field available to create the foundation of group technology.

The final approach to classification and coding "nomenclature" is effectively utilized by the Asian Rim. Here, the name of the part becomes the first major differentiator. The problem with the English language as we use it is that a washer and a spacer could be the same part, could have exactly the same geometry, but could have no relationship in terms of their name.

As Wickham Skinner and others have pointed out, the notion of group technology is absolutely fundamental to the concept of focused factories and the foundation on which those focused factories are built. Skinner and others have dictated we must create manufacturing cells that are groupings of manufacturing resources to produce a family of products. The cell can be a logical cell, which means we simply identify the pieces of equipment and may not have them be physically close to one another; or it can be a physical cell, where dissimilar pieces of machinery are brought together to form a part family. The fabrication process then has the ability to produce significant savings in time and cost.

An example, as recognized in the late 1970s and early 1980s by Larry Burner and Ed Turner at Rockwell International, demonstrates. In the production of waveguide, Turner and Burner recognized that it took 23 moves and 17 weeks to produce a piece of waveguide for the microwave radios Rockwell was producing in Dallas. The production flow before the application of group technology appeared to be a spaghetti bowl. But with the conscious application of group technology, it now takes only 9 moves and 1.7 weeks to create the same part family. This is shown in Chapter 3.

In summary, the notion of group technology is a fundamental prerequisite to the recognition that similarities exist. Where standardization opportunities apply, both flexibility and efficiencies can be accrued in an integrated process design and development environment.

## SECTION 4.3 TOTAL QUALITY MANAGEMENT

Just as group technology forces us to think about similarities, total quality management (TQM) forces us to think about the *process of doing things*. As stated by the Department of Defense, TQM is a:

> DoD initiative for continuously improving its performance at every level, in every area of DoD responsibility. Improvement is directed at satisfying such broad goals as cost, quality, schedule, and mission need and suitability. TQM combines fundamental management techniques, existing improvement efforts, and specialized technical skills under a rigorous, disciplined structure focused on continuously improving all DoD processes. It demands commitment and discipline. It relies on people and involves everyone.[12]

Although this reads as though it is only for the government, what has evolved in the world is the recognition that *it is not only the end result that warrants review, but it is the process by which the team got to the result!*

Continuous improvement is based on a simple principle that models what the "as is" condition is of every process and then develops a strategy and tactics for sustainable improvement. The notion is laden with methodologies such as modeling and characterization. Briefly, methodologies for modeling allow a team of cross-functional experts to "draw a picture" of the process in a way that the process can be analyzed. This could be a factory-floor machining process, a setup process, or a purchasing process. Any time actions solicit a result and affect the overall customer satisfaction, they should be analyzed. Methodologies for characterization involve rigorous and controlled testing that can allow a cross-functional team of users to understand fully how a process achieves a result, not just what the result is. The Department of Defense initiative contains four elements that should be in every TQM plan:

1. A vision—long-term focus on the direction of the organization that recognizes the challenges of change and competition.
2. Principles—fundamental concepts that serve as the basic rules for management decisions and actions.
3. Practices—plans that implement the principles and demonstrate and reinforce the desired TQM behavior through systematic and continuous improvement applications.
4. Techniques—technical methods employed that provide systematic, specific road maps for accomplishing specific tasks.

## SECTION 4.4 MATERIAL MANAGEMENT IMPLICATIONS FOR INTEGRATED PROCESS DESIGN

As companies migrate from a supply-push to demand-pull environment, Figure 4–7 applies. Supply push has previously focused on looking behind to see what parts have been in the queue, and then demand strategies for pulling parts from the queue have been utilized to optimize the capacity within a particular operational element of the factory.

**FIGURE 4–7**
**A Comparison of Supply Push with Demand Pull**

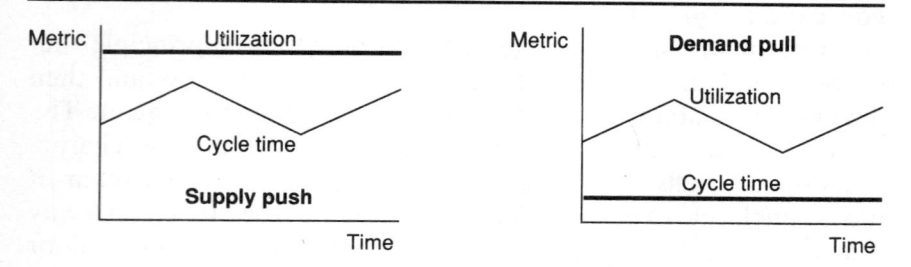

**Resolve:**
Clearly define the corporate metrics for world-class competition.
*Time* is becoming the new corporate metric.

---

This strategy says make what you can and has created significant inventory, bottlenecks, and cycle times in many American factories. The migration from supply push to demand pull has been a nontrivial transition because the metrics for demand pull are significantly different from the metrics for supply push. In essence, demand pull looks ahead and asks what does my customer need? (customer is defined as the next person in line or the next process in line) and how can I best satisfy that customer's requirement? By optimizing inventory and cycle time, the demand pull environment can support the customer but at the expense of utilization, which in this case will vary. The transition from utilization optimization to cycle-time optimization, from utilization maximization to cycle-time minimization is a fundamental shift that must affect the integrated process design and development environment.

The implications of the transition from supply push to demand pull must also be well understood as we migrate to integrated process design and development because, as Shigeo Shingo and others present,[13] we must understand what the material management philosophies are and how they affect the design tooling, the fixturing, and the implications of the process, and how rapidly we must set up and tear down products and processes to optimize the overall production flow. A summary of the two thoughts follows:

*Supply push*—optimize the utilization of the machines and the people.

*Demand pull*—optimize the cycle-time flow of materials.

As we progress in our understanding of world-class competition, we find several flaws in the systems that formally develop material management in today's supply-push environment. If our task is to respond very rapidly to changing customer demands, then the formal schedules often are not valid given today's flexible environment. Purchasing and manufacturing priorities differ, capacity planning is poor and very difficult to perceive and assign correctly, the constant firefighting mode is present, and the system is relatively inefficient, ineffective, and costly to operate. Configuration management is very difficult because the engineering change control is difficult to manage.

As we migrate to the 21st century, we must resolve these significant management problems with simplified systems and simplified process design. We must provide significant visibility and measurement techniques to provide correct capacity and cycle times as they reflect what the customer truly needs. Yield factors and scrap must be measured, and continuous improvement action plans must be put in place with targets that are met. Systems must be integrated, and bases of information must be created that can be utilized for many purposes but in consistent and logical fashions. Procedures must be disciplined, and bill of materials must be accurate to the six sigma level of quality. Finally, routings and standards for the particular process must be flexible in the face of change and yet provide rigorous and adequate tracking methodologies to ensure that we know how every part is produced. The customer-integrated requirements must be met.

The material management philosophy of the 21st century has a significant impact on integrated process design. Just-in-time demand pull requires synchronization and simplification. The Kanban concept has significantly different metrics than today's traditional capacity utilization. Finally, demand pull requires significant changes in teamwork, visibility, communication, commonality of product family, ability to identify commonality of product family, simplicity of operation, singularity in terms of master scheduling at the end of the line, worker security with no inventory security blanket, and significant consistency of purpose.

This migration to material management of the 21st century in a just-in-time environment requires significant culture shifts, just as in concurrent engineering and group technology. It requires significant investment in education and simplification.

## SECTION 4.5   PROCESS CAPABILITY DESIGN

By approaching the area of process capability design from a core competencies viewpoint, new and novel answers are evolving in this important area. By understanding the technological food chain shown in Figure 4-8, a new approach to process capability design can be developed. This figure depicts the hierarchy of technology, from the equipment supplier to the raw material and subcomponent supplier to the final product producer. Those of us that buy television sets buy from the final product producer.

In the case of high definition television (HDTV), the secondary gains in importance. The HDTV concept is heavily linked to the ability of the semiconductor industry to supply the rapidly responding chip sets needed for digital picture processing. But the chip sets are only as good as the equipment available to produce the

**FIGURE 4-8**
**The Technological Food Chain**

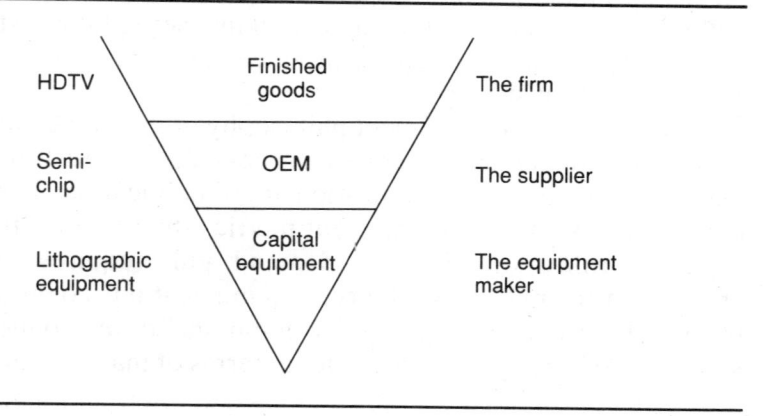

| | | |
|---|---|---|
| HDTV | Finished goods | The firm |
| Semi-chip | OEM | The supplier |
| Lithographic equipment | Capital equipment | The equipment maker |

fine line widths needed to move the electrons on the semiconductor. In the past, the equipment market and the subcontract and materials markets have been viewed as secondary marketplaces that had small market sizes. They fell into a "niche" market category from Michael Porter's analysis of how strategies for products can be developed.[14] However, as the United States is finding with HDTV, the final product is dependent on the subcontracted semiconductor chip set, which is dependent on the equipment supplier to provide the capability for the desired process. In this light, the secondary marketplace takes on a much more strategic nature. Hence, Porter's model is changing as we begin to focus on time and integration.

Key concepts such as "core competencies" espoused by Prahalad and Hamel[15] and strategic alliances are now forcing the "art" of tactical process capability design into the "science" of strategic process capability design. Although the common theme in all of this is to focus the company's attention on what it is capable of performing as the best in the world, the notion of alliances has become of paramount importance as one understands the above concept of the technological food chain. Clearly, there must be an in-house competency for all strategic aspects of the business—even if the supplier provides the talent and the knowledge to execute the technology. Whether it be suppliers of equipment or suppliers of materials and subcomponents, the issues of core competencies and strategic alliances have forced this topic to become more strategic than ever before.

## SECTION 4.6 BENCHMARKING

"Benchmarking is the search for those best practices that will lead, to the superior performance of a company. Establishing operating targets based on the best possible industry practices is a critical component in the success of every business."[16] To be successful in global competition requires knowledge of the customer as well as knowledge of the competition. Robert Camp at Xerox and others have found that benchmarking forces the organization to truly identify the critical success factors needed to be competitive. As shown in Figure 4–9, the benchmarking process is made up of five detailed sets of steps:

**FIGURE 4-9**
**Benchmarking Process Steps**

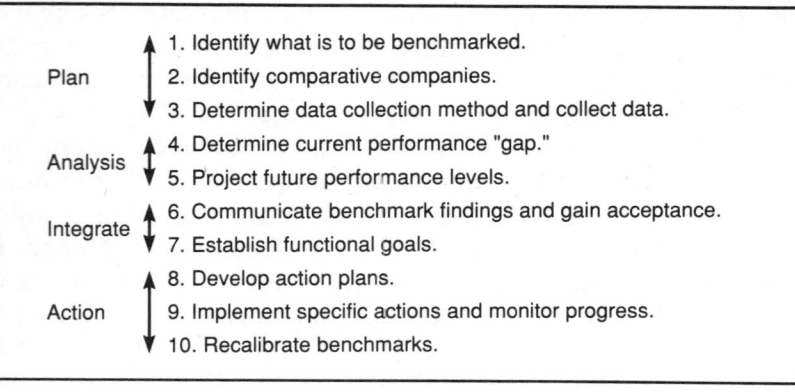

| | |
|---|---|
| Plan | 1. Identify what is to be benchmarked. |
| | 2. Identify comparative companies. |
| | 3. Determine data collection method and collect data. |
| Analysis | 4. Determine current performance "gap." |
| | 5. Project future performance levels. |
| Integrate | 6. Communicate benchmark findings and gain acceptance. |
| | 7. Establish functional goals. |
| Action | 8. Develop action plans. |
| | 9. Implement specific actions and monitor progress. |
| | 10. Recalibrate benchmarks. |

1. Plan the benchmarking process.
2. Analyze the data and determine the "gap."
3. Integrate the results and establish the goals.
4. Develop action plans.
5. Monitor these actions until leadership is attained.

Camp's text on benchmarking goes through each of the 10 steps in detail. Three issues must be addressed as an organization launches a benchmarking exercise. The first is that the "best in class" for a particular function may not be a competitor. For example, if one of the core competencies for global competition is to establish a world-class distribution system, comparisons with Federal Express or Sears, Roebuck & Co. may offer more insight into the overall process than analysis of the distribution channels of the competition.

Second, do not assume the benchmarking data are readily available. Operating performance data can be derived from the annual reports of the companies in question. Facts such as sales per employee, research and development budgets, and globalization trends can be found. However, facts on time to market, yields, or quality data are much more difficult to find. Unless one of these is deemed useful to capture market share, these performance metrics are not often published.

Third, the notion of benchmarking should not be used as a target so much as a yardstick. During the time it takes to analyze the benchmark situation and go through Camp's 10 steps to meet the goal, the company that set the benchmark has made progress as well. Clear goals that will lead to superior performance are mandatory for the successful usage of benchmarks.

## SECTION 4.7 MATERIAL HANDLING AND MATERIAL FLOW SYSTEMS IMPLICATIONS FOR INTEGRATED PROCESS DESIGN

Designing an integrated system requires three fundamental elements:

1. Integrated information systems.
2. Integrated control systems.
3. Integrated material flow systems.

These three fundamental elements of any integrated manufacturing system will be reflected in Section 4.10 as we introduce a methodology for modeling these three approaches. Suffice it to say here that integrated material flow is fundamental to any integrated systems design. There are many ways to approach material flow systems for a 21st century manufacturing environment.

Dr. Jim Tompkins writes in *Industrial Engineering* that two fundamental elements of material flow must be understood to achieve integration.[17] The first is material flow within the cell. The cell here referenced is the group technology cell shown earlier in Section 4.2. Material flow within the cell must be in a streamlined flow, typically in a U-shape, allowing for simplification and integration and ultimately for automation, if automation makes sense. It moves product through in a symmetrical fashion and allows for ease of movement of parts within the cell. Material flow between cells uses the foundation of what Tompkins calls the "spine concept," where the spine is the integral material handling system to link multiple cells on a straight line with cells flowing from the periphery, as shown in Figure 4–10. The spine concept in Tompkins' model allows for flexibility of flow within each cell and also expan-

**FIGURE 4–10**
**The "Spine Concept"**

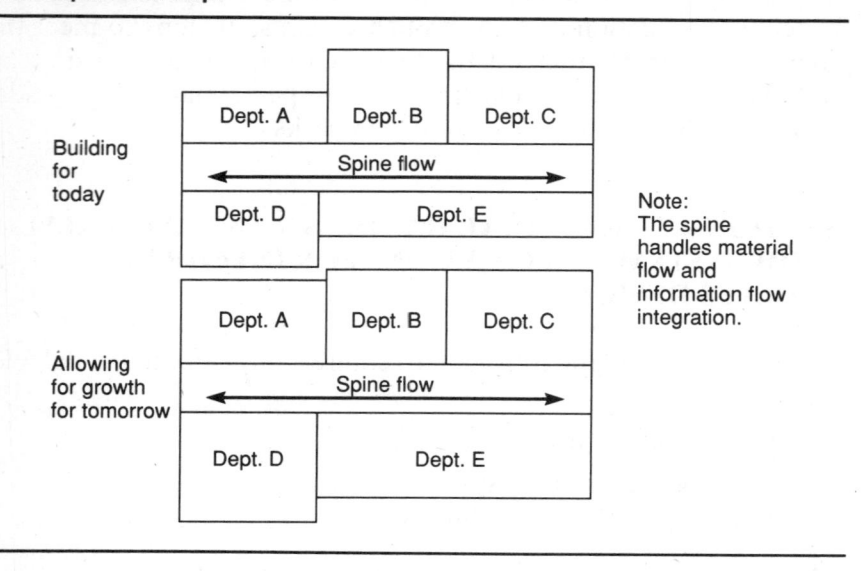

```
                    ┌─────────┬─────────┬─────────┐
                    │ Dept. A │ Dept. B │ Dept. C │
   Building         ├─────────┴─────────┴─────────┤
   for              │    ←──── Spine flow ────→   │
   today            ├─────────┬───────────────────┤      Note:
                    │ Dept. D │      Dept. E      │      The spine
                    └─────────┴───────────────────┘      handles material
                                                         flow and
                    ┌─────────┬─────────┬─────────┐      information flow
                    │ Dept. A │ Dept. B │ Dept. C │      integration.
   Allowing         ├─────────┴─────────┴─────────┤
   for growth       │    ←──── Spine flow ────→   │
   for tomorrow     ├─────────┬───────────────────┤
                    │ Dept. D │      Dept. E      │
                    └─────────┴───────────────────┘
```

sion capacity and capability within each cell while maintaining flexibility and integration of material flow between the cells.

To achieve automation of the material handling and material flow systems requires understanding several necessary steps. First of all, we must ask, is the task necessary? Has the simplification exercise been conducted to such an extent that we have truly identified a value-added process? Second, is the material flow streamlined? Has a U-shaped type flow been instituted? And is this material handling automation system justified? Usually, through simplification and streamlined flow, much savings have already accrued, and automation is difficult to justify on a cost basis. However, in many cases in today's world-class manufacturing environments, automation is being justified on a quality basis and a time basis rather than cost.

Finally, are the individual tasks simplified to the point where automation can be achieved at its simplest level? Concepts such as automated guided vehicles (AGVs), automated storage and retrieval systems (AS/RSs), smart pallets, and others create this foundation for the automated material handling system of the 21st century. Automated guided vehicles are, in essence, delivery trucks moving at a fixed pace throughout a factory picking up and dropping off items in a programmed manner. These AGVs perform the intercell material flow in a programmed manner. The spine, in an intercell flow pattern, often is an automatic storage and retrieval system. The AS/RS holds inventory in a fixed location where, with identifiable pallet locations in a high bay environment, automation takes parts to and from these pallet locations and delivers them to the next cell in line. Finally, smart pallets are being utilized in integrated material handling systems in order to identify where the product is with respect to the various locations within the factory.

All of these cases require significant data tracking mechanisms. To achieve integrated material flow, it is fundamental to know exactly where every product is. Today's cost-accounting systems that assess the financial value of the inventory have zero value when it comes to finding a part. Data integrity at material levels requires piece part counts by location where the exact status of each part is known. A multitude of methodologies are being employed in material flow systems to perform this data tracking. What is needed is a flexible methodology that can track and yet provide accurate and reliable feedback in terms of where products are located. Bar coding has come to the forefront in today's environment for automatically tracking products, tooling, equipment, material handling devices, smart pallets, and so on. Most of these methods are made up of simple blocks of lines linked in some sort of coding format such that as they pass a scanning mechanism, a bar code can be read and interpreted and the code for that particular product or process can be tracked to that location.

## SECTION 4.8  SETUP AND TOOLING STRATEGIES INCLUDING "SINGLE MINUTE EXCHANGE OF DIES"

In J. T. Black's text, *The Design of the Factory with a Future*,[18] he outlines an approach to establish a world-class factory design. In this approach, the first steps review the layout and the setup methodologies needed for good process integration with the facility. To supplement Black's methodology, the following initial steps are recommended for the overall integrated process for facilitization and tooling strategies:

1. Simplify—make the parts and the processes as standard and as simple as possible
2. Lay out the facility according to the general, strategic guidelines of Tompkins
3. Assess the setup strategies according to the general, strategic guidelines of Shigeo Shingo outlined in Section 4.8.
4. Establish a total quality strategy according to the TQM philosophy outlined in Section 4.3.
5. Establish an integrated, preventative maintenance strategy.
6. Integrate the production planning and control aspects of material management with the facility.

We now focus on the need for simplified setup as a major aspect of integrated process design and development.

The concept of the single minute exchange of dies (SMED) was initiated by Shigeo Shingo.[19] In his text, Shingo focuses process integration on the inventory of a company. In the past, inventory was perceived as an asset to most companies. But with *time* as the new competitive metric for customer satisfaction, inventory now is being viewed as a liability in many cases. It forces larger lot sizes, it often is of the wrong mix, and it causes huge expenses in material variances when engineering changes are required. Shingo noted the need for inventory was an indirect result of the inability of the production facility to respond rapidly to changes in demand and design. This was a direct result of the inability of the production facility to achieve a setup time of zero time units.

The product mix can be of any type, but the primary focus here shall be on low-volume, high-mix types of facilities. And although this sounds like a job shop arrangement, this concept of low volume, high mix is exactly the target of "customer responsiveness" no matter how many of each product we wish to produce. Low volume, high mix requires many setups and teardowns. One approach to fix the problem is to lump common setups together so that more value-added time can be achieved from each piece of capital equipment. This will be successful only at the detriment of increased inventories and inefficient use of production planning and control techniques. A second approach is to use much more knowledgeable employees to operate the machines than are otherwise needed because they have become "setup engineers" when complex jobs are undertaken. Again, although the direction of the enterprise is to have more multidisciplined workers, it may be a costly solution.

The recommendation by Shingo is to not get caught in the economic order quantity "blind spot" of assuming that setup times are fixed. Figure 4–11 shows the classic EOQ configuration. By comparing the setup time effects (P) with the inventory carrying costs (S), the intersection has been classically used for the calculation of the EOQ. At point (E), the setup versus inventory aspects balance and this is perceived as the best quantity. But it assumes that the setup times (P) are fixed.

**FIGURE 4–11**
**The Economic Lot Size Graph**

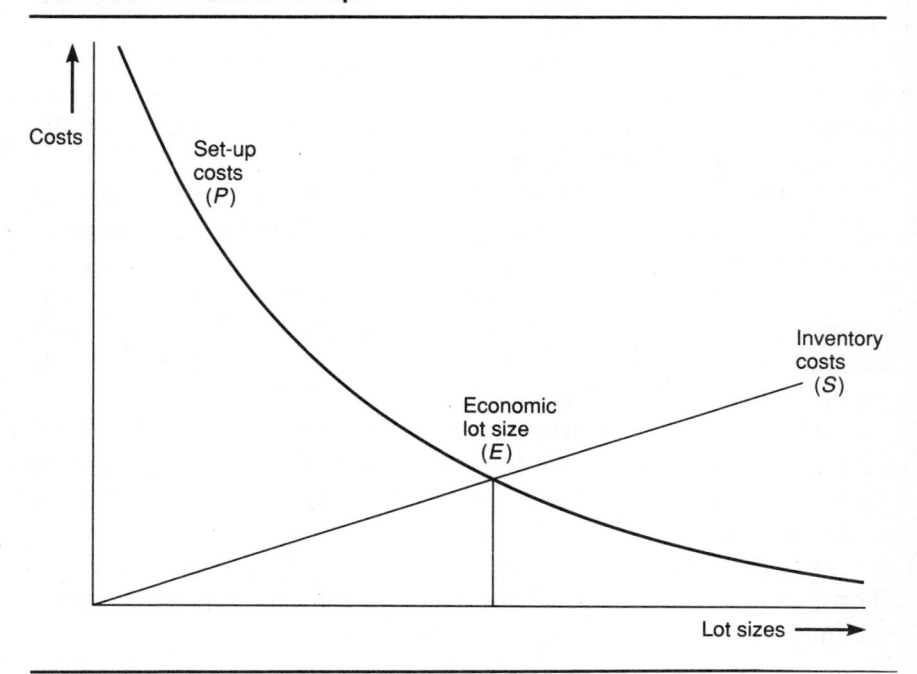

In Bedworth and Bailey,[20] this is calculated using the following, classic formula:

$$EOQ = \sqrt{\frac{2 \times D \times P}{I + S}}$$

where:   D is the demand in units per year.
P is the cost of purchase or cost of setup in dollars per batch.
I is the interest charges in cost per unit per time period.
S is the charge for storage in cost per unit per time period.

If setup could be reduced to a very small value, then the lot size could also be reduced and the responsiveness to changing customer demands would not be hurt by inventory valuations.

The approach provided by Shingo to attack SMED is divided into four steps. These are:

**1.** Phase 0—Preliminary stage internal and external setup conditions are not distinguished. Through informal observations and discussions with the operators, 80 percent of the setup time improvement opportunities can be identified.

**2.** Phase 1—Separating internal and external setup. Internal setup refers to the actual machine adjustments and other aspects of setup that consume valuable machine hours. External setup refers to those activities that can be performed off-line from the machine so that production time can be optimized. This is the critical step because doing as much as possible externally will optimize the value-added aspects of the machinery.

**3.** Phase 2—Converting internal to external setup. Clearly, if internal setup costs production time while external setup can be done off-line and in parallel, the real objective is to maximize the external setup.

**4.** Phase 3—Streamlining all aspects of the setup operation. Through detailed analysis of the remaining internal setup activities, optimization of these times can be achieved as well.

## SECTION 4.9  SYSTEM INTEGRATION DESIGN PHILOSOPHY

In every case referenced in this chapter, the prerequisites for integrated process design have built on the notion that simplification is fundamental to the overall success of the system. Heretofore, engineers have prided themselves on the ability to design complex systems that required sophisticated tooling, sophisticated fixturing, sophisticated controls, and sophisticated management systems to perform the operation properly. This pride of creation has yielded very, very complex systems, and the cost-effectiveness of these complex systems is being questioned. The model for the 21st century appears to be one of simplification first, followed by integration and, ultimately, automation.

The stepwise process shown in Figure 4-12 is espoused by the author. It, in essence, says that for any system integration design to be successful, first we must educate all parties involved. Then we go through a rigorous simplification process. Then we educate again in terms of what the integration requirements are, and then we develop the integration plan. Ultimately, we educate again and finally, we consider automation only where automation makes sense.

In today's environment, most savings have come from simplification, with integration and automation coming in second and third. However, as stated earlier in this chapter, automation does have a place if used to justify quality of effort or timeliness of effort.

**FIGURE 4-12**
**The Modern Systems Analysis Approach**

      Educate
       Simplify
        Educate
         Integrate
          Educate
           Automate
            Educate!

## SECTION 4.10  INTRODUCTION TO AN INTEGRATED PROCESS DESIGN AND DEVELOPMENT SYSTEMS DEFINITION METHODOLOGY

We offer here a simple methodology as a step to understanding how the integrated process design and development system must evolve. The basis for this methodology is called the *ICAM Def*inition language (IDEF), a public domain technique put into place by the U.S. Air Force Integrated Computer Aided Manufacturing program in the late 1970s. The extension of this, called the "IDEF0/Triple Diagonal," or IDEF0/td,[21] developed by the author, was put in place in the mid-1980s as a simplifying approach based on the recognition that planning a system from the top down may never get to the point of recognizing how a part actually flows through the process. It is also based on the premise that the cross-functional team must have a consistent definition vehicle that allows all members of the team to visualize how the system operates today (as is) and view how the system can operate tomorrow (to be).

The triple diagonal is an extension of the IDEF methodology, which builds a model bottom up based on simplifying assumptions found by the National Institute of Standards and Technology (NIST) in its work on the Advanced Manufacturing Research Facility (AMRF). NIST found that only three types of functions are performed: planning functions, control functions, and execution functions. This is consistent with the recognition that integrated systems evolve from integrated information, integrated control, and integrated material and tool flow. Information relates to planning, control to control, and material and tool flow relates to the execution diagonal in this model.

However, the triple diagonal can be used for more than simply modeling material flow systems. It can be used to model any process by which some function is being performed. The acquisition of a part through the purchasing function, the design of a product through the design function, the cutting of a purchase order through the purchasing function, all these functions can be modeled with the triple diagonal concept. The notion is very simple. In essence, it says, to understand the system we must first understand how basic execution functions, often "material" or "documents," flow through the process.

The first step is to understand the as is environment and rationalize how best products are going to flow through the overall process. The model for this triple diagonal is based on the syntax that is shown in Figure 4–13a where the function is modeled as a block diagram using a "verb-noun" phrase, inputs come in from the left, outputs go out the right, controls come in from the top, and mechanisms come in from the bottom. Inputs are those elements of the enterprise that are transformed by the functions to create the output. Controls are those elements of the enterprise that, when the function is performed, will still exist. Mechanisms are the means by which the functions are performed.

For example, if we are performing a machining operation, the function would be "perform machining" (see Figure 4–13b). The input would be the raw stock; the output would be the machined part. The control would be the numerical control (NC) tape and the schedule; the mechanism would be the machine tool and/or the operator. In this simplified example, we could model anything. The function might be "design product," "perform stress analysis," or "create purchase order." In all of these, inputs are transformed to outputs; controls develop strategies for how to perform the functions; and mechanisms are the means by which those functions are employed. The triple diagonal is simply building on the linkage of each of these functions in a structured manner that has three lines of functions shown in Figure 4–14. In this figure, one can see that

**FIGURE 4–13a**
**The IDEF0 Model Syntax**

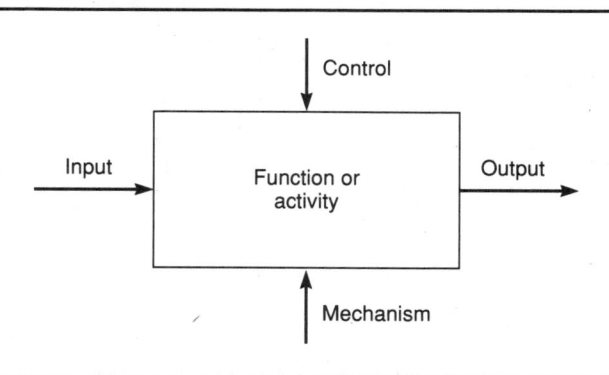

**FIGURE 4–13b
An IDEF0 Model Example**

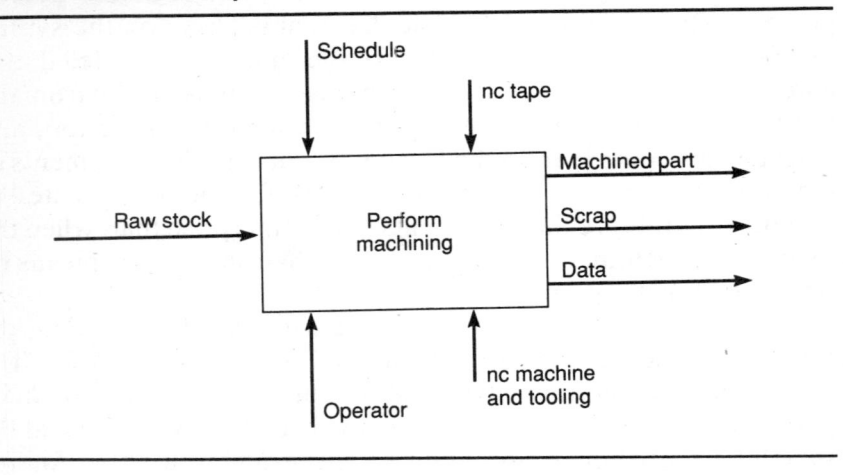

```
                    Schedule
                            nc tape
                      ↓       ↓
                    ┌─────────────┐   Machined part
                    │             │ ─────────────→
       Raw stock    │   Perform   │   Scrap
      ────────────→ │  machining  │ ─────────────→
                    │             │   Data
                    └─────────────┘ ─────────────→
                      ↑       ↑
                            nc machine
                            and tooling
                    Operator
```

**FIGURE 4–14
An IDEF0/td Model**

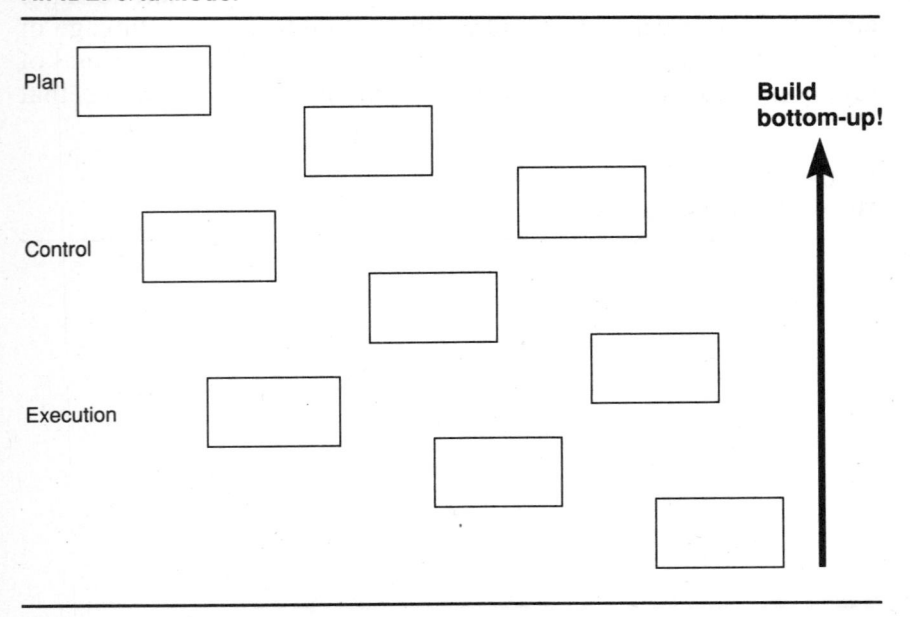

Plan

Control

Execution

**Build
bottom-up!**
↑

planning functions form a diagonal, control functions form a diagonal, and execution functions form a diagonal.

We begin by determining the basic flow of the product or the functions. Step one determines this basic flow and is modeled first to establish the context (see Figure 4–15a). Step two adds the basic material flow between these functions (see Figure 4–15b). Hence, as we look at the result after step two, we have a basic production flow model that the shop floor assembly operators, the shop floor foremen, and all process planners can understand as they relate to how a product is actually made. Step three adds the control feed-forward mechanisms (see Figure 4–15c). This model then represents the second level in the diagonal where controls of inventory, controls of the machine tools, and all other controls must be understood. Step four closes the loop. As we have stated earlier, to become world class we must have closed-loop feed-forward and feedback systems in every instance (see Figure 4–15d). The closed-

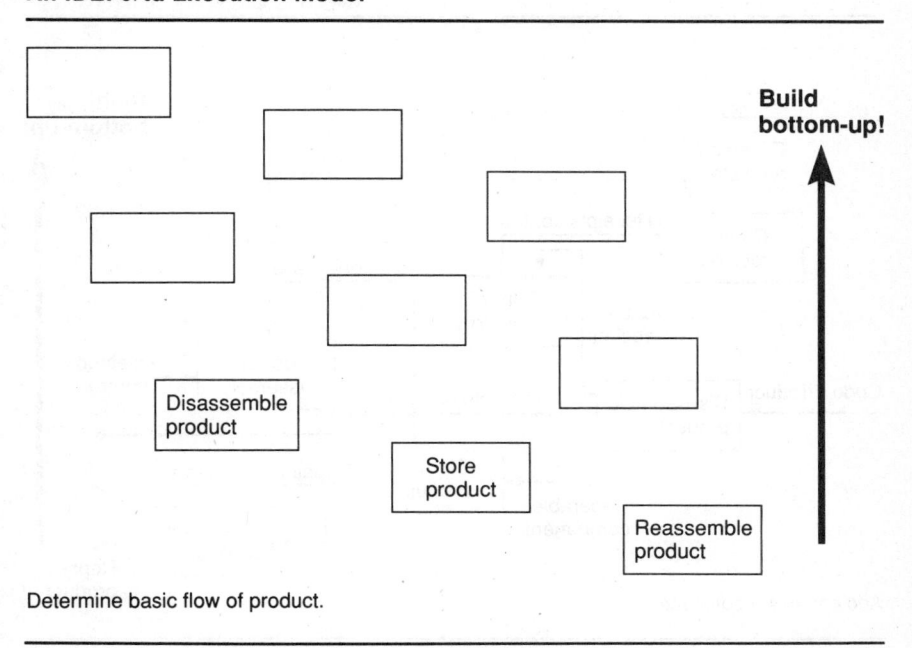

**FIGURE 4–15a**
**An IDEF0/td Execution Model**

Build bottom-up!

Disassemble product

Store product

Reassemble product

Determine basic flow of product.

**FIGURE 4–15b**
**An IDEF0/td Execution Model with Material Flow**

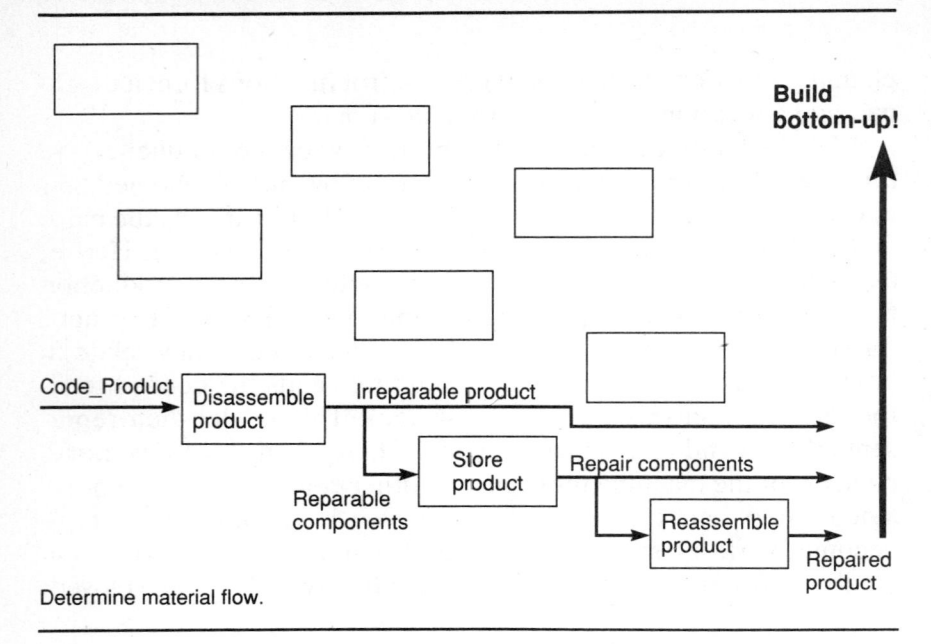

Code_Product → Disassemble product → Irreparable product →
Disassemble product → Reparable components → Store product → Repair components →
Store product → Reassemble product → Repaired product

**Build bottom-up!**

Determine material flow.

**FIGURE 4–15c**
**An IDEF0/td Control Model**

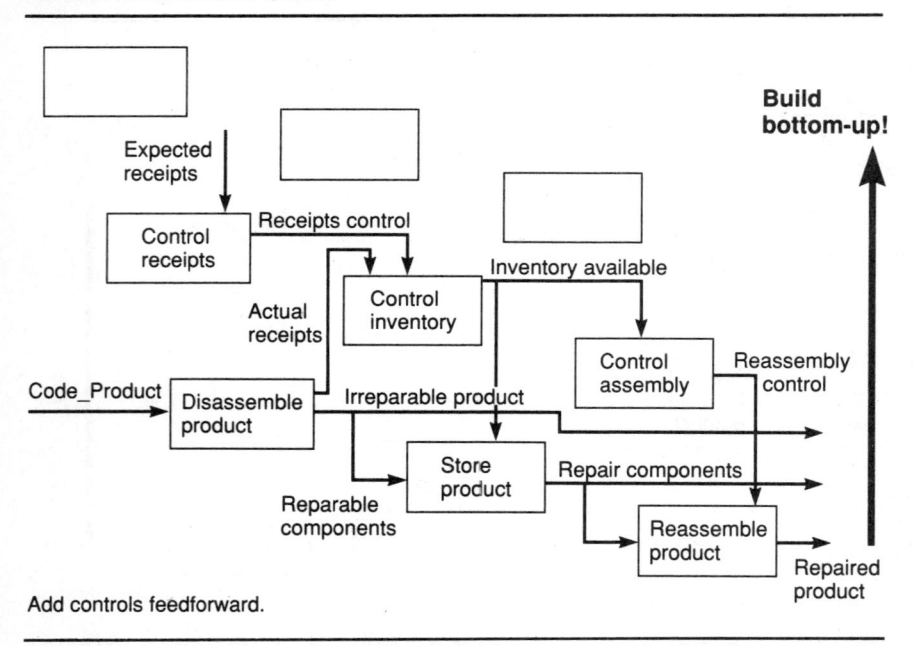

Expected receipts → Control receipts → Receipts control
Control receipts → Actual receipts → Control inventory
Control inventory → Inventory available → Control assembly → Reassembly control
Code_Product → Disassemble product → Irreparable product
Disassemble product → Reparable components → Store product → Repair components
Store product → Reassemble product → Repaired product

**Build bottom-up!**

Add controls feedforward.

74

## FIGURE 4–15d
### An IDEF0/td Control Model with Feedback

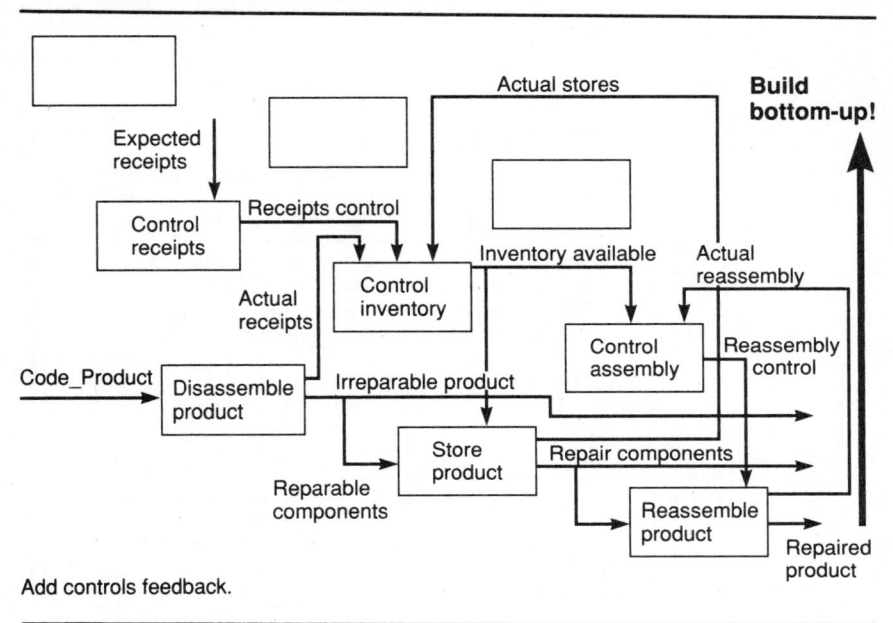

*Add controls feedback.*

loop feedback systems from the execution diagonal back to the control diagonal is fundamental to maintaining control of the factory floor.

Step five adds the planning functions (see Figure 4–15e). What functions are involved in the overall definition and development of a product? Step six closes the loop again, one of the closed-loop functions from control back to planning that allow us to operate in a closed-loop environment (see Figure 4–15f). Step seven begins to populate the model at the execution level by quantifying and simulating the material flow and adding elements such as the time to perform a process, the quantity each function is able to produce, the head count required to perform that process, and so on (see Figure 4–15g). Finally, step eight is the quantification and simulation of the information flows necessary to perform the process (see Figure 4–15h). These typically are represented at the planning and control levels within the triple diagonal architecture.

**FIGURE 4–15e**
**An IDEF0/td Planning Model**

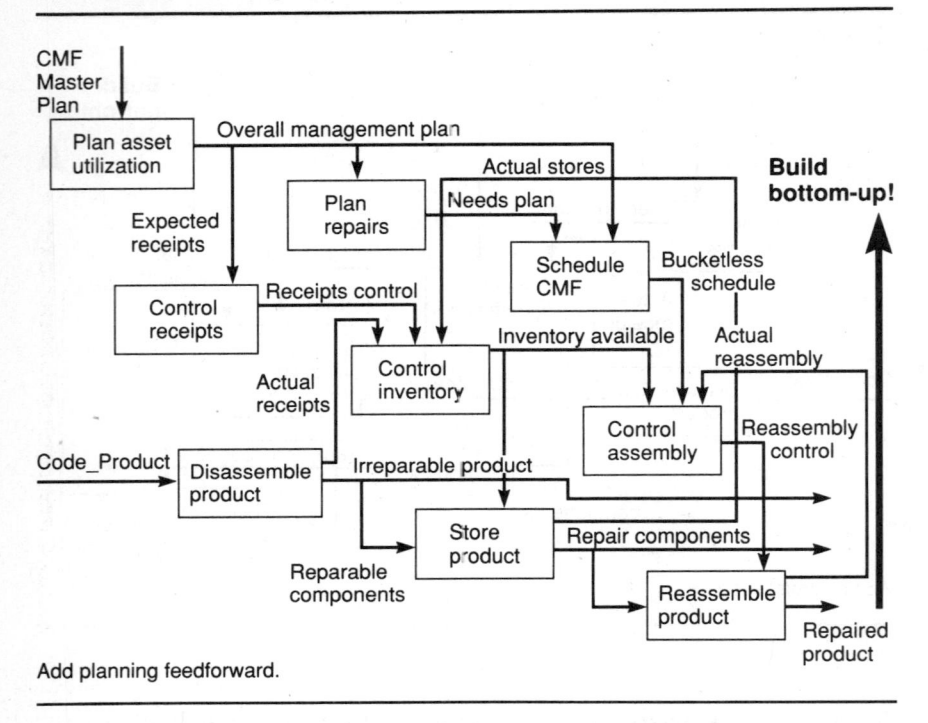

```
CMF Master Plan
   │
   ▼
┌──────────────┐  Overall management plan
│ Plan asset   ├──────────────────────────────────┐     Build
│ utilization  │           Actual stores          │     bottom-up!
└──────┬───────┘                                  │
       │        ┌──────────┐ Needs plan           │
  Expected      │  Plan    ├──────────┐           │
  receipts      │  repairs │          │           │
       │        └──────────┘     ┌────┴─────┐ Bucketless
       ▼                         │ Schedule │ schedule
   ┌──────────┐ Receipts control │   CMF    ├──────┐
   │ Control  ├──────────┐       └────┬─────┘      │
   │ receipts │          │  Inventory available    │ Actual
   └──────────┘          ▼            │            │ reassembly
         Actual      ┌──────────┐     │     ┌──────┴────┐
         receipts    │ Control  │     │     │  Control  │ Reassembly
                     │ inventory│     └────▶│  assembly │ control
                     └──────────┘           └───────────┘
Code_Product  ┌────────────┐ Irreparable product
──────────────▶│Disassemble ├──────────┐
              │  product   │          │
              └────────────┘          ▼
                  │              ┌─────────┐ Repair components
             Reparable           │  Store  ├─────────────┐
             components          │ product │             │
                                 └─────────┘        ┌────┴──────┐
                                                    │ Reassemble│
                                                    │  product  │
                                                    └───────────┘
                                                         │
                                                    Repaired
                                                    product
```

Add planning feedforward.

The development of the td model shown progressively in Figure 4–15 has proved to be very useful. The model shown is a derivative of a model used in a repair facility. It was created by a team of users and systems developers in a few weeks and effectively used by the team to visualize how a repair facility, complete with probabilistic incoming bill of materials, can migrate from a supply-push asset management plan to a demand-pull asset management plan. The team collectively came to the conclusion after much deliberation centered on the model.

In summary, we have presented here a methodology that will be utilized throughout the rest of this text in a manner that demonstrates how planning integrated process design and development systems and controlling integrated process design and development systems can lead to an integrated process design and development environment.

**FIGURE 4–15f**
**An IDEF0/td Planning Model with Feedback**

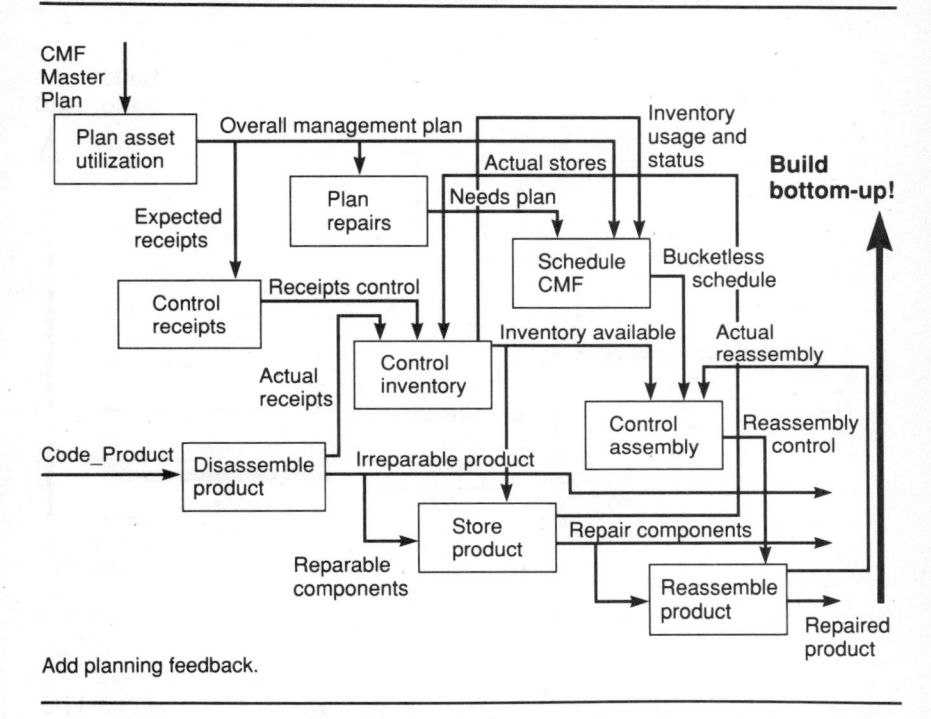

Add planning feedback.

## SECTION 4.11   SIMULATION AND ITS ROLE IN INTEGRATED PROCESS DESIGN AND DEVELOPMENT

The IDEF0/td methodology shown above can be extended into the direct support of a simulation model of the design and development system. In the past, PERT and CPM charts[22] have been successfully used to monitor and control major projects. These have certain limitations, though, in that both tend to use deterministic values in the process times and move times. Hence, a critical path can be determined, but with little real feel for the confidence interval or the sensitivity to changes in certain functions. By employing the IDEF0/td methodology, we have shown how a function model can be created that mirrors the process flow at the execution level and the information flow at the planning and control levels. These two

**FIGURE 4–15g**
**An IDEF0/td Quantified Material Flow Diagonal**

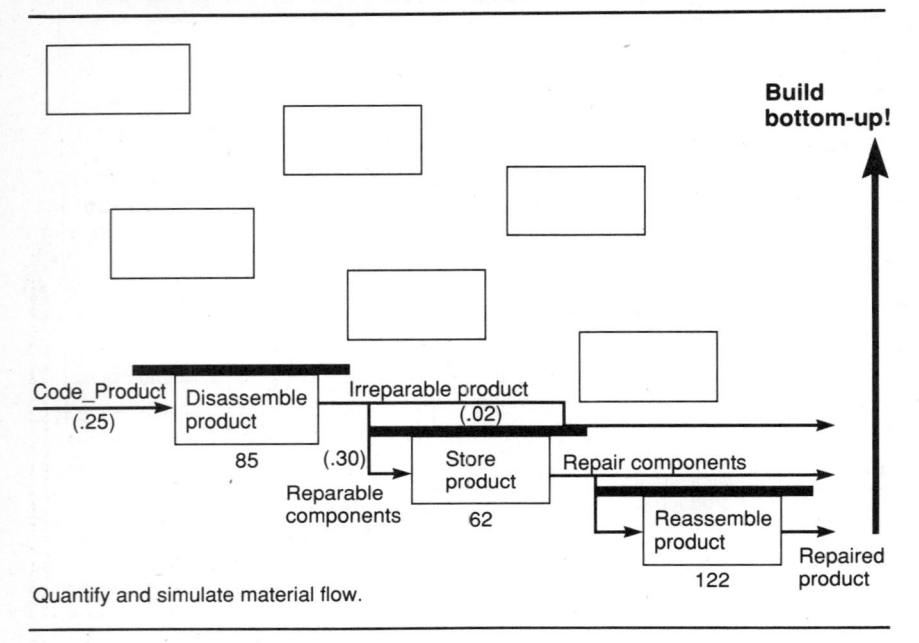

Code_Product | Disassemble product | Irreparable product (.02)
(.25) | 85 | (.30) | Store product 62 | Repair components | Reassemble product 122 | Repaired product

**Build bottom-up!**

Reparable components

Quantify and simulate material flow.

models are of extreme importance to any project manager. The simulation of the execution diagonal will predict a confidence interval for the overall process creation. The simulation of the planning and control diagonal will depict how much traffic will be on the information systems needed for feed-forward and feedback. Questions such as the following can be routinely answered through a simulation model:

1. What is the critical path, and with what confidence?
2. What are the bottleneck operations, and how sensitive are they?
3. What is the predicted manufacturing throughput time, given a predefined sequencing algorithm and predefined materials management algorithm?
4. What if we reduce the inventory levels?

**FIGURE 4-15h**
**An IDEF0/td Quantified Control and Planning Diagonal**

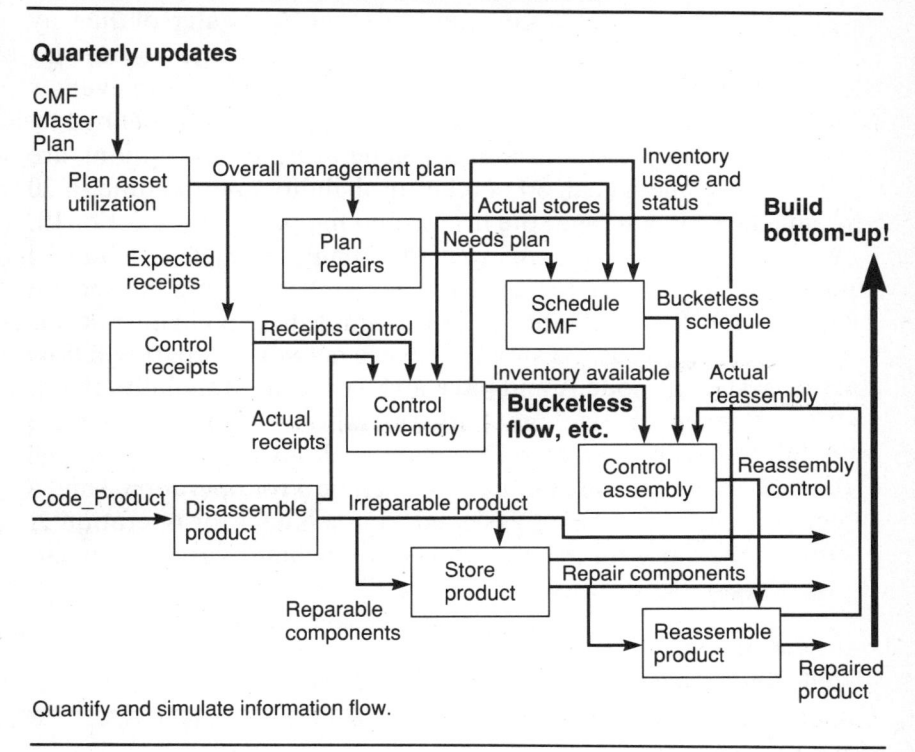

**Quarterly updates**

[Diagram: Flow chart showing quarterly updates process with the following elements:
- CMF Master Plan → Plan asset utilization
- Overall management plan → Plan repairs
- Actual stores, Needs plan → Schedule CMF
- Inventory usage and status
- **Build bottom-up!**
- Expected receipts → Control receipts
- Receipts control
- Actual receipts → Control inventory
- **Bucketless flow, etc.**
- Inventory available → Bucketless schedule
- Actual reassembly → Control assembly → Reassembly control
- Code_Product → Disassemble product
- Irreparable product → Store product → Repair components
- Reparable components → Reassemble product → Repaired product]

Quantify and simulate information flow.

5. What if we change the material management philosophy from supply push to demand pull?

All of these questions are asked daily by good program managers, and few can be answered unless a simulation model has been created.

Simulation models work in a simple way. With discrete event simulation, each event (or in our case, function) is defined with an expected time and a time variability. These functions are then interconnected to form a network. Simulations can be created for such things as bank teller workstations, computer-aided design scheduling, or process routing predictions. Any time a model can be created with defined time averages and variations, a simulation can be built. Then, using a technique called "Monte Carlo" methodology, time is simulated by taking a slice of time every step of the way (e.g., a slice may be every minute or every hour and so on). By keeping track of how each event is progressing and when an event is scheduled to end, the decision as to what happens next is defined by probabilities and a random number generator. If two options are available, and one has an 80 percent probability and the other a 20 percent, then, on average, the first would happen 8 times out of 10. However, this assumes there are many, many samples. In a limited sample, the first may only happen 5, 6, 7, or 9 times and the second the remainder. Let us create a simple example to explain the notion.

Figure 4-16 creates a simple process network. A part will flow into the system at node A and have an 80 percent probability that it will be assigned to machine 1 for operation 1 and a 20 percent probability it will be assigned to machine 2. After this, it has a 60 percent chance of being assigned to machine 3 for operation 2 and a 40 percent chance of being assigned to machine 4 for operation 2. Assuming the times for the operations by machines vary somewhat,

**FIGURE 4-16**
**A Monte Carlo Simulation Model for Process Flow**

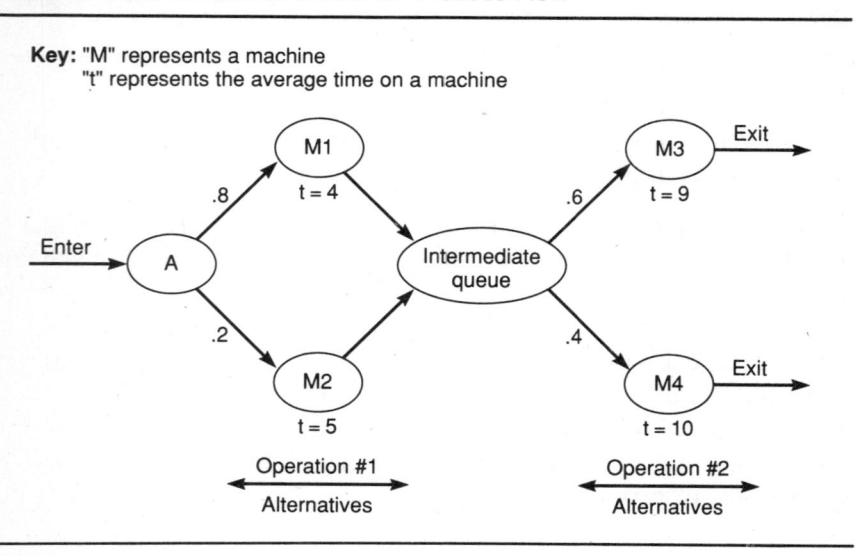

what is the expected manufacturing throughput time for this scenario?

One way to calculate the answer is to calculate the *deterministic* expected value. For this it would be as follows:

Expected value = .8(Time M1) + .2(Time M2) + .6(Time M3) + .4 (Time M4)

So, the result would be

.8(4) + .2(5) + .6(9) + .4(10) = *12.7 time units*

The first thing you notice is that the expected value is in tenths of a time unit when all we can have are whole units. The second is that this is the expected value if this were done many, many times. Now let us use a random number-generated Monte Carlo simulator and assess what is the expected value and the variability of these values. To do this, we shall generate a random number string of digits from zero to nine. This stream can be found in any mathematics, probability, or statistics book. Our stream shall be:[23]

10 49 00 15 02 96 24 68 12 72 52 50 61 29 99 23 33 21 46 59

This represents 20 pairs of "two-ples" and can represent each pair of decisions that a part must use for its traversal through the factory. If we allow the first number in each set to represent the decision on whether we select machine 1 or 2, we can use the rule that if the first digit in our pair is a 0–7, this represents 80 percent of our sample, and it will therefore go to machine 1. But if the first number is an 8 or 9, this represents 20 percent of our sample; hence, the part goes to machine 2. The same will be used for the second machine selection. The result is as follows:

| Part Number | Stream | Sequence | Total Time |
|---|---|---|---|
| 1 | 10 | M1–M3 | 4 + 9 = 13 |
| 2 | 49 | M1–M4 | 4 + 10 = 14 |
| 3 | 00 | M1–M3 | 4 + 9 = 13 |
| 4 | 15 | M1–M3 | 4 + 9 = 13 |
| 5 | 02 | M1–M3 | 4 + 9 = 13 |
| 6 | 96 | M2–M4 | 5 + 10 = 15 |
| 7 | 24 | M1–M3 | 4 + 9 = 13 |
| 8 | 68 | M1–M4 | 4 + 10 = 14 |
| 9 | 12 | M1–M3 | 4 + 9 = 13 |

| Part Number | Stream | Sequence | Total Time |
|---|---|---|---|
| 10 | 72 | M1–M3 | 4 + 9 = 13 |
| 11 | 52 | M1–M3 | 4 + 9 = 13 |
| 12 | 50 | M1–M3 | 4 + 9 = 13 |
| 13 | 61 | M1–M3 | 4 + 9 = 13 |
| 14 | 29 | M1–M4 | 4 + 10 = 14 |
| 15 | 99 | M2–M4 | 5 + 10 = 15 |
| 16 | 23 | M1–M3 | 4 + 9 = 13 |
| 17 | 33 | M1–M3 | 4 + 9 = 13 |
| 18 | 21 | M1–M3 | 4 + 9 = 13 |
| 19 | 46 | M1–M4 | 4 + 10 = 14 |
| 20 | 59 | M1–M4 | 4 + 10 = 14 |

The average time for the 20 parts is total time of 269 time units divided by the 20 parts equals 13.45. But this does not agree with our expected value of 12.7 time units. The reason is that the simulation represents only a small sample size. Given 1,000 parts, it would more closely come to the expected value; given 1 million, it would be very close. But most of our factories are running small lots with high mix to respond to the customer. If this is true, then the variability of this shop mixture must be known. The next sample of 20 parts might yield a sample average of 12.5 and so on.

As we learn how factories must operate in the 21st century, the knowledge of the mean is certainly not enough. It is the variability that must be managed, and any algorithmic calculation is not going to provide the visibility into this variability that a simulation can provide. Thus, the essence of simulation.

## REFERENCES AND RECOMMENDED READINGS

1. W. Edwards Deming, *Out of Crisis* (Cambridge, Mass.: MIT Press, 1986).
2. James L. Nevins and Daniel Whitney, *Concurrent Design of Products and Processes* (New York: McGraw-Hill, 1989).
3. James L. Bossert, *Quality Function Deployment* (Milwaukee, Wis.: ASQC Press, 1991).
4. "A Smarter Way to Manufacture," *Business Week,* April 30, 1990.
5. Kimberley Beaumariage and Dan Shunk, "Issues in Migrating to Teamwork," CASA Blue Book Series, Society of Manufacturing Engineers, 1991.

6. Charles Savage, *Fifth Generation Management* (Boston: Digital Press, 1990).
7. Geoffrey Boothroyd and P. Dewhurst, *Design for Assembly Handbook* (Kingston, R.I.: Boothroyd and Dewhurst Associates, 1985).
8. Bart Huthwaite, *Bart's Laws* (Rochester, Minn.: Institute for Competitive Design, 1987).
9. Wickham Skinner, "Manufacturing—Missing Link in Corporate Strategy," *Harvard Business Review,* May–June 1969.
10. Alex Houtzeel and C. S. Brown, "A Management Overview of Group Technology," CASA/SME Westec Conference, March 1982.
11. John L. Burbidge, *Production Flow Analysis* (Turin, Italy: Turin International Center, 1969).
12. *Total Quality Management* (Washington, D.C.: Department of Defense Publication, 1988).
13. Shigeo Shingo, *A Revolution in Manufacturing: The SMED System* (Stamford, Conn.: Productivity Press, 1985).
14. Michael Porter, *Competitive Strategy—Techniques for Analyzing Industries and Competitors* (New York: Free Press, MacMillan, 1980).
15. C. K. Prahalad and G. Hamel, "The Core Competence of the Corporation," *Harvard Business Review,* May–June 1990.
16. Robert C. Camp, *Benchmarking* (Milwaukee, Wis.: Quality Press, American Society for Quality Control, 1989).
17. James Tompkins, "The SPINE Concept," *Industrial Engineering,* March 1983.
18. J. T. Black, *The Design of the Factory with a Future* (New York: McGraw-Hill, 1991).
19. Shingo, *A Revolution in Manufacturing.*
20. David D. Bedworth and James Bailey, *Integrated Production Control Systems* (New York: John Wiley & Sons, 1987).
21. Dan L. Shunk, Jerry Cahill, and William Sullivan, "Making the Most of IDEF Modeling—The Triple Diagonal Concept," *CIM Review,* Fall 1986.
22. Frank S. Budnick, Richard Mojena, and Thomas Vollman, *Principles of Operations Research for Management* (Homewood, Ill.: Richard D. Irwin Inc., 1977).
23. Ibid.

# CHAPTER 5
# PLANNING THE PROCESS FOR INTEGRATED PROCESS DESIGN AND DEVELOPMENT

## SECTION 5.1   PERFORMING A TECHNOLOGY ASSESSMENT

As we've seen, the theory of manufacturing is changing substantially. As Drucker pointed out, "We cannot build it yet, but already we can begin to specify the post-modern factory of 1999." He states there will be four fundamental issues: First, statistical quality control will become paramount with information and accountability aligned and rigorous and reliable feedback provided. Second, in new manufacturing accounting systems the cost of direct labor will become much smaller than in the traditional sense, and we must identify not only the costs of production but also the costs of nonproduction. Third, we will migrate to a modular manufacturing environment—a potentially global, modular manufacturing environment—where design may be performed in one location, planning in a second location, actual fabrication in a third location, and possibly assembly at a fourth location. These locations may be next door or on the other side of the world. This flotilla of modules must be linked through an information network. Fourth, we will take a systems approach that imbeds the physical process into the overall economic process of the environment. To perform this technology assessment for process design and development, we must begin by recognizing where the technology has been and where the technology is going.

To compete anywhere in the world, we must be able to produce a lower-cost product with much higher quality in much less time and truly meet the real needs of the customer. To do so, the technology has changed from developing product design and then transitioning in a serial fashion to process design. In a market-driven orientation, the "engineered product" is the business of the future. Many companies throughout the world are developing ways by which new manufacturing technologies can produce new manufacturing capabilities. An engineered product is the unique product configuration that appears like a totally custom product to the customer. Yet the engineered product fits neatly into a group technology-based process family when viewed internally to the design and manufacturing organization. This unique set of characteristics allows the enterprise to meet the needs of the customer yet maintain efficient operations inside. These complementary concepts affect the changing markets and allow companies to develop new ways to compete in these world marketplaces.

This manufacturing strategy is forcing new organizations and new management styles. The slang for this is to be able to produce a "custom product for the customer and yet make it look as though it's a 'jelly bean' manufacturing environment for the factory." In essence, the customer has unique requirements. The unique requirements must be served in a very efficient and timely manner for companies to be successful in product and process development. To do so requires a modular manufacturing environment based on certain technological principles, such as group technology, that can be very efficient at what it does and yet flexible enough to create a custom product.

The technological assessment to plan the process must begin with true understanding of where the technology is going and where customer requirements are driving us. In this regard, the ability to produce an economic order quantity of one is a necessary condition for success. This is not to say that order quantities of one will be produced on a regular basis, but the ability to track, plan, tool and position these parts, and produce economically one unique part at a time is fundamental to the overall philosophy espoused in this text. For this to be successful, the technology assessment and all other aspects of the business must be well understood, recognizing that to be successful the customer's requirements must be met in a very timely manner. This is the essence of Figure 5–1.

## SECTION 5.2  BENCHMARKING THE INTEGRATED PROCESS

Competitive benchmarking has clearly become a major aspect of technology assessment. By comparing the systems and the procedures of the enterprise with others that have similar functional

**FIGURE 5–1**
**Focus on the Planning Functions**

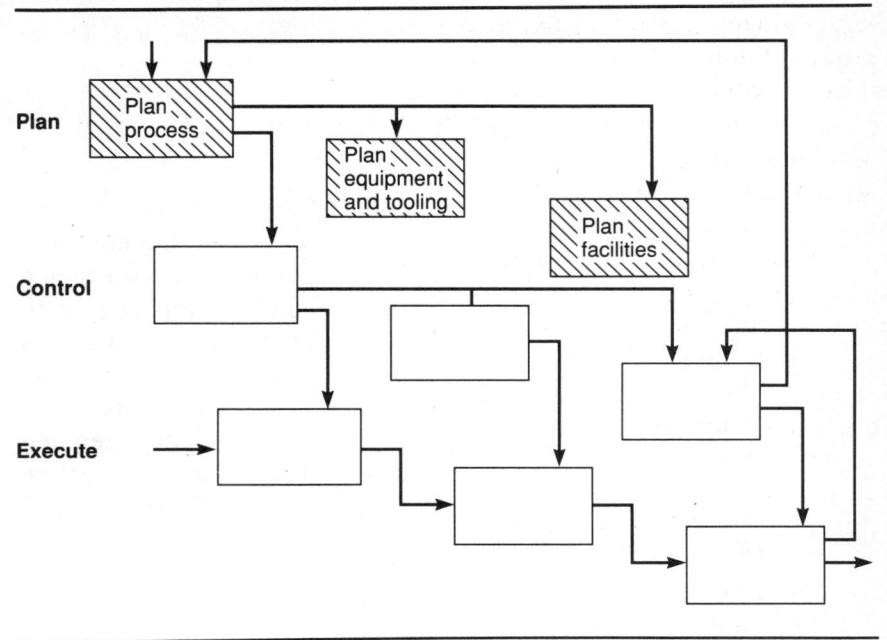

requirements, a truer vision can be created that targets world-class achievements. Results from Xerox[1] and other companies keen on benchmarking have shown that the comparison need not always be against the competition. For example, if the enterprise must distribute the product to the customer in a very timely fashion, a comparison of the distribution practices of the enterprise with Federal Express or DHL may be just as revealing as analysis of the competitor's practices. The guideline for benchmarking is to *compare against the best!*

Competitive benchmarking has been simplified here to a four-step, closed-loop process. The steps are:

1. Plan the analysis.
2. Conduct the analysis.

3. Understand the results of the analysis.
4. Create the action plan and monitor the results.

For each of these steps, certain guidelines are recommended. These follow:

**1. Plan the Analysis**

To plan the analysis, the functional model of the enterprise must be created and analyzed. This functional analysis should highlight the strategic functions of the enterprise where world-class performance must be present. As we stated earlier, if distribution is key to success, then plan the analysis around the best in class. The plan should capture the strategic thrust of the organization and should identify which criteria will be measured and how the measurements will be conducted. For example, if quality of the product is deemed critical, is this a function of quality of the process, and if so, then how can you capture process quality data from the "best in class" organizations?

**2. Conduct the Analysis**

The successful way to conduct the analysis is through the creation of a cross-functional, virtual task team that may be chartered with worldwide visibility. This team should be constituted with sufficient balance that the results will be viewed as credible and without bias. The approach to conducting the benchmarking exercise begins with a thorough literature search on the criteria in question. For example, if a key criteria is customer awareness of the product, often a public library assessment of the advertising budget for the best-in-class companies is a valuable first step. This will not provide all of the information, but it will provide initial insight into the strategy of the company in question.

Openly announcing your intentions for a benchmarking analysis to the company in question is proving to be more successful than clandestine efforts to capture data. Most world-class companies recognize that it is the corporate culture and the execution of the plan that allows them to be world class rather than a secret strategy. Motorola was the first Malcolm Baldrige award winner. Every month Motorola hosts a massive seminar for suppliers and customers that outlines how it was able to instantiate the six sigma corporate philosophy. Wallace Engineering in Houston, the 1991 Baldrige winner, does the same for small businesses. These are excellent launching points for conducting the analysis.

### 3. Understand and Interpret the Results

Don't get to this step and then ask, what does all of this mean! With a rigorous first step that defines what is really critical for success and with a well-designed plan, understanding the results should be easy. The interpretation is a bit more difficult because this forces the cross-functional team to interpret the best-in-class results and then overlay the philosophies found onto the functional model of the enterprise in order to determine what actions are needed for success.

### 4. Create the Action Plan and Monitor the Results

The cross-functional team is chartered to create the action plan. This must clearly show how the critical aspects of their enterprise can be strategically improved by the actions recommended. For example, Wallace Engineering reports it needed to become more responsive to its customer. It found that no procedures for handling customer requests were present, thus presenting the recipient of a customer call with no guidelines on how to handle the call. With a rigorous process modeling and improvement program, Wallace now has a plan to handle any customer request in one hour or the customer is allowed to define the resolve. This is world class! But it took a team of motivated people to recognize that rigor needed to be a strategic thrust of the company.

Once the plan is in place and actions begin to happen, the team must continue to monitor the results in a closed-loop fashion that reflects on the overall planning needs identified in step one. Each 3 or 6 or 12 months, the team should review progress toward attaining world class. Remember, the benchmark you establish is changing over time. To use the benchmark as a way to focus attention on areas to improve is admirable, but to use the benchmark as the target of the company guarantees that the company will always be trying to catch up to someone. It is an excellent way to determine where you are—not the answer to where you should be.

## SECTION 5.3 ACHIEVING FLEXIBILITY DUE TO PRODUCT DEFINITION UNCERTAINTIES

When designing and implementing manufacturing systems that operate at the leading edge of strategy and technology, time to develop software and hardware elements for the factory becomes excessively long.[2] As Figure 5-2 depicts, as the life cycle of the product is drawn on a time line, the knowledge about the design tends to lag the knowledge required to design and build the manufacturing systems necessary for production.

Productivity improvements and potential strategic alliances with equipment and material suppliers are driven by the knowledge available early in the process. An approach to this paradox is to design *flexible manufacturing systems* (FMS) that can support all parts and assemblies within a predefined, strategic part family boundary. But the issue with FMS in the past has been the lack of "F" in the FMS. Although flexibility is critical, an FMS purchased by a major American manufacturer of farm equipment in the early 1980s had the following scenario play out:

**FIGURE 5-2**
**Product Definition Paradox**

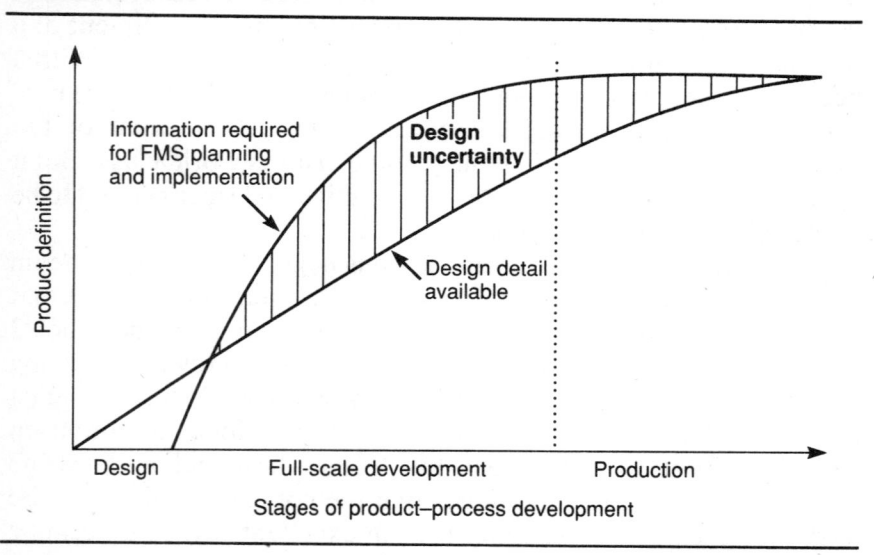

1. The FMS was purchased to be installed at location A and designed to machine a complete family of more than 100 parts.
2. After the order was placed but before shipment, the mission of the FMS was changed, and it was redirected to location B.
3. At location B, the FMS was asked to machine only five parts.
4. Once in operation, the projected savings never materialized.
5. After less than one year in operation, it was sold to a competitor at 30 percent of the purchase price.

The system was deemed a failure and, in the years that followed, few were even planned. But the system was not meant to be a transfer line, it was designed for flexibility. Figure 5–3 shows how

**FIGURE 5–3**
**Comparison of Lot Size with Cost per Piece**

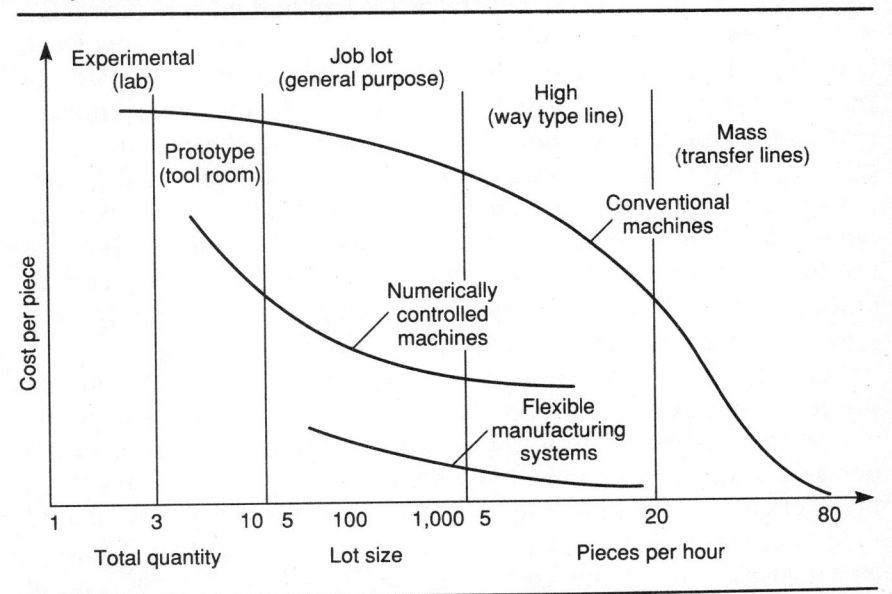

these strategies begin to evolve for the creation and assessment of process technology. In Figure 5-3, lot size is compared to cost of the product. As lot size increases and cost decreases, new strategies for process must be used. At the lower right-hand corner, the cost per piece is very small because the conventional machines had been configured into a mass production, transfer line. But the technological push for the 21st century is to keep the cost per piece low yet move to the left on the lot size axis. Notice that a technological gap still prevents producing a lot size of 1 to 10 total pieces for only pennies.

## SECTION 5.4  DEFINING THE PROCESS NEEDS FOR A PRODUCT

To define the process needs, we begin with the basis by which we develop our three examples; that is, the traditional product definition characteristics of form, fit, and function defined through a rigorous QFD methodology.[3] This will be utilized to translate the customer requirements into the process requirements. The design team and the manufacturing team must create the product-process strategy early in the product life cycle (as highlighted in Figure 5-2).

In today's manufacturing environment, these two teams work concurrently at a roundtable environment early in the design process. Why? Because data have shown that by the end of conceptual and preliminary design, 80 to 90 percent of the product's life cycle cost has been defined. Dr. Gene Merchant displayed the chart in Figure 5-4 at a conference as early as the 1960s where he pointed out that as we progress through the various life cycle elements of the product—conceptual design, preliminary design, detail design, full-scale development, and ultimately fabrication, assembly, and support of the product—we encounter two types of costs.

The first is the incurred cost of actual dollars expended on that product. And as is clearly pointed out in the diagram, these actual costs often are very, very low as we exit the design stage. In the design stage, we may only have spent 2 to 5 percent of the total product cost by the time it leaves that phase. However, the second type of cost is the determined or commitment of the costs. Clearly, 80 to 90 percent of the costs are dictated by the time we leave

**FIGURE 5–4
Actual versus Incurred Cost Comparison**

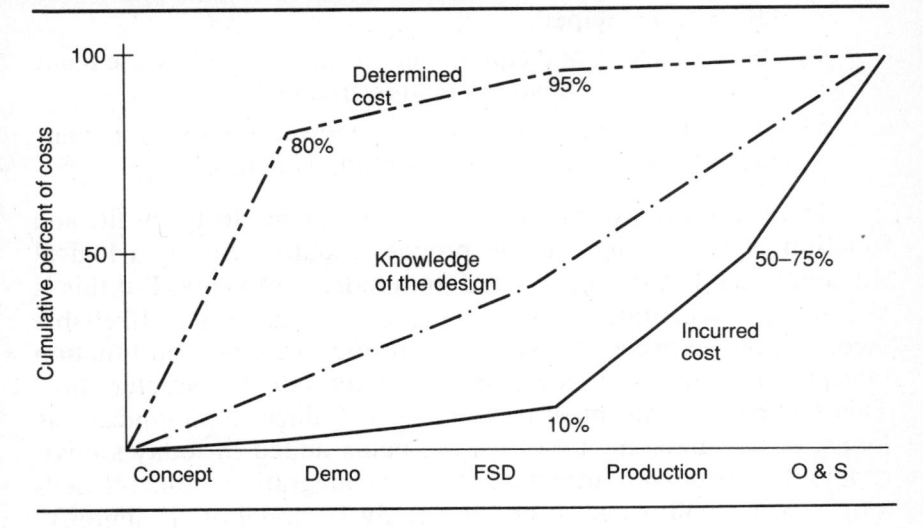

[Graph showing Cumulative percent of costs vs. project phases (Concept, Demo, FSD, Production, O & S). Curves shown: Determined cost (reaching 80% by Demo, 95% by FSD), Knowledge of the design, and Incurred cost (10% at FSD, 50–75% at Production).]

design, leaving very little leverage in terms of the overall efficiencies that can be accrued if we do this planning in a serial fashion. The notion is to be able to provide rigorous and reliable feedback from the planning information bank back to the design concepts as early in the design phase as possible, such that rigorous and reliable feedback can be achieved. It is imperative to have an integrated information network in place to carry this out.

Product needs and process needs are becoming integrally linked. In successful world-class manufacturing, these two fundamental notions are done in a timely manner as early in the design stage as is technically feasible. The steps being recognized today are as follows:

1. Quality function deployment and other tools are employed to define what the customer requirements are and the life cycle needs.
2. An internal product-process team is chartered to compare the needs to the product features and define the form, fit, and function of these features.

3. As concurrently as is technologically feasible, the product-process team determines the baseline by which the process design is developed.
4. Costs, budgets, and equipment are all specified by the team and a pro-forma business plan is created.
5. The entire planning process is reviewed with senior management and a "go–no go" decision is made.

The translation from customer requirements to form, fit, and function of the product to the process requirements is a logical transition that has been exploited for hundreds of years. But this is becoming much, much faster, and there is rigorous and reliable feedback of the process design back to the form, fit, and function interpretation and, ultimately, back to the product characteristics. This feedback, with integrated and well-defined communications loops, is the fundamental transition being added in today's environment. Through simplification and integration, this closed-loop process will occur. Using the analysis model of (1) simplify, (2) integrate, then (3) automate, it may or may not be sufficient to go only to step two for success. But the successful automated systems have all progressed through steps one and two before launching the automation activities.

## SECTION 5.5  CREATION OF VARIOUS TYPES OF PROTOTYPES

As Jan Drees, retired senior design representative from the Bell Helicopter Company, pointed out in a National Research Council panel report on rapid prototyping recently, "There are various ways to create a new product."[4] Often, it begins with a new idea that germinates in a variety of ways. There is no forced entry into the creation of a new idea. Many times the new idea must go through what is called a "proof-of-concept" phase in order to create a research base by which the idea can be tested in a variety of formats. As the new idea matures and progresses to the point where companies wish to introduce the idea to the customer base, a spectrum of ways is available to develop these new products. The spectrum ranges from the ability to produce the first product and sell it to the ability to require significant, intense prototyping—in order to develop this product. Figure 5-5 attempts to capture Jan Drees' comments. You can see that the mature technology evolution to the emerging technology is a fundamental spectrum.

If we are dealing with a mature process that is characterized and understood, then no prototyping is required in today's environment. If we are dealing with a product development that has primarily mature technologies but a bit of emerging technologies, then preproduction prototyping may be adequate. However, if the technologies to be developed for a process are emerging or it's a new and a high-risk environment, then experimental and intense prototyping are often recommended to develop a new product.

Today, decreasing the time it takes to go through this prototyping stage is emphasized to rapidly get a product to market. Through the creation of "virtual product environments," the developer of a prototype can learn about the product and/or the process before committing to the final product and/or process. "Virtual environment" means that the company has characterized its products and its processes to such a level of sophistication that it can simulate the product-process integrated environment without building any prototypes. Today, many new approaches are being utilized to achieve rapid prototyping.

**FIGURE 5-5**
**The Spectrum of Prototype Development**

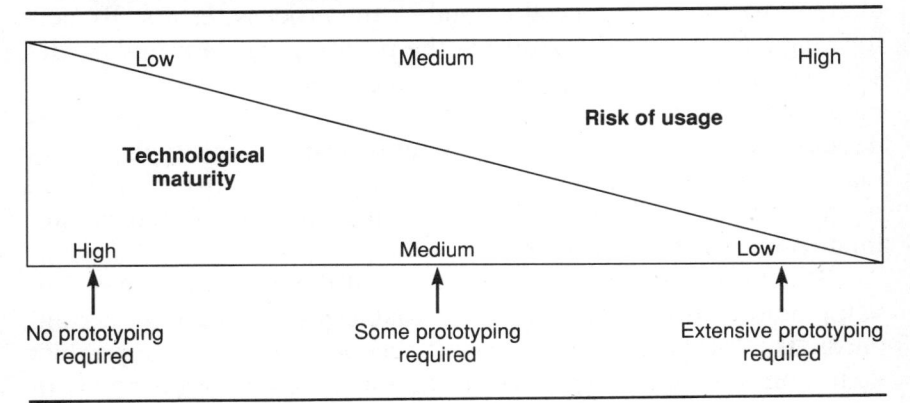

During the National Research Council deliberations on rapid prototyping, four definitions of prototyping were developed. These are:

**1.** *Proof of concept.* The initial experimental model is developed that facilitates basic demonstration and testing of the new product or feature concept. This uses the fastest route available just to demonstrate what the concept is.

**2.** *Proof of product.* In this type of prototyping, the exact form of the design to test the product's functionality and expectations is developed. The method by which this is manufactured has little bearing on this phase of the prototyping process.

**3.** *Proof of producibility.* Here, a more sophisticated approach to prototyping is developed that integrates not only product design but also the specifications with the manufacturing process capabilities to determine mass production feasibility. This is a true test of all of the "ilities" that are necessary to support a product—reliability, maintainability, serviceability, supportability, sustainability, and so on.

**4.** *Proof of production.* Here, finished products are used to introduce and refine new production techniques to ensure selected materials conform and to identify potential production bottlenecks.

These four phases of prototyping must be clear in the minds of all who are developing new product-process development cycles in order to exploit and understand where best to invest in new methodologies to create various prototyping processes. Gathering information in a structured manner during prototyping allows the organization to learn about the process in such a skillful manner that a characterization of the product and/or the process results. By performing structured testing during the prototyping stage and by developing a rigorous design of the experiment allows the investment in prototyping to pay dividends later for other product-process developments. Through characterization, some prototyping steps may be eliminated. Moreover, a "good" product may be feasible in the first try—a remarkable event today that can allow for significant time-to-market compression.

It is also fundamental at this stage of the product-process development to utilize the "best" available product definition standards. By "best" is meant the latest accepted, most robust product definition standard accepted by the national or international body of standards makers that affect your business. In many companies today, the standard for simple product definition is called "IGES," the initial graphics exchange specification (or standard). IGES is a standard way to define the geometry of a product and transport that geometry between multiple computer-aided design (CAD) systems to achieve some element of transportability.

IGES was created from the needs specified by the CAD system suppliers. In the late 1970s, many CAD systems were evolving, but none used the same protocol to define a geometry. Once a product was defined on one system, it would have to be redefined if it were ever ported to a second system. The U.S. Air Force ICAM (integrated computer-aided manufacturing) program in conjunction with the National Bureau of Standards and individuals from Boeing and General Electric launched the first IGES initiative to develop a common geometry definition. The initial offering and its embellishments lasted much of the 1980s, even though it was severely limited by handling only geometry. But remember, it was billed as the *initial* graphics exchange specification. It was never expected to become a world standard because there were standards bodies throughout the world working on a robust system to do this. In the mid-1980s, the International Standards Organization (ISO), in conjunction with the American National Standards Institute (ANSI), the U.S. Department of Defense, and a multitude of companies developed an international standard for product definition that may meet the needs of the world in the 1990s. In the United States, this was called the "product definition exchange specification," or PDES, while the Europeans launched a similar activity called "STEP." In 1991, these two thrusts were merging into what may be the needed worldwide standard. With the evolution of PDES using STEP, a robust means to define not only the product geometry but also the topology, the materials, and so on can be used as a consistent means to transport valuable product definition data throughout the world.

## SECTION 5.6  CREATION OF THE PROCESS PLAN

Today, the process plan is typically developed through characterization of the part and then translation of that part requirement to equipment assignments and then stored in a fixed format with preassigned times and anticipated capacities. An example of a process plan creation will clarify this procedure. The source of the following steps is the state-of-the-art survey of computer-aided process planning conducted by Alting and Zhang.[5] Figure 5–6 depicts the classic 10 steps to create a process plan.

The first step is interpreting the design data. This is often in blueprint form, but a digitized definition of the product is now available. Critical elements of data must be appended to the geometry. These are:

- Expected product quantities and life cycle.
- Expected batch size.
- Geometry of the finished product as well as of the raw material stock used to begin the process.
- Raw material properties.
- Dimensions and tolerances.
- Surface finish.
- Any required heat treatment or hardness criteria.

Notice that in this classic creation of the process plan, the geometry has already been specified and the batch sizing recommended. In the integrated process design and development scenario that we are creating in this text, the process design experts would participate in these decisions early in the design stages.

**FIGURE 5–6**
**The 10 Classic Steps to Create a Process Plan**

1. Interpretation of product design data
2. Selection of machining processes
3. Selection of machine tools
4. Determination of fixtures and datum surfaces
5. Sequencing of operations
6. Selection of inspection devices
7. Determination of production tolerances
8. Determination of the proper cutting conditions
9. Calculation of the overall times
10. Generating of process sheets including NC data

Source: Leo Alting and Hongchao Zhang, "Computer-Aided Process Planning—A State of the Art Survey," *International Journal of Production Research* 27 (1989).

The second step involves translating the geometry to process requirements. And the actual machining operations are specified (e.g., mill, drill, grind, bore, etc.). With the notion of integrated process design, we will review a shift in this second step in Chapters 9 and 10. The plan for the process may be generated 6 to 60 months in advance of the product being produced. To integrate this in a timely manner and to accommodate a worldwide supplier base consideration, the second step is now evolving into two substeps:

1. Define the process requirements.
2. Define the alternative machining operations available.

The third step assigns the actual machining center for each process step in order to assess capacity requirements and overall productivity measures.

The fourth step analyzes the product at a detailed processing perspective, including control surfaces and fixturing strategies.

The fifth step depicts the "best" sequence of operations necessary for a successful, quality part to be produced. The correct sequence of milling operations or drilling operations must be defined to cost-effectively produce the product. Alting states this is often a corporate strategy based on the capabilities of focus at the specific corporation and the capabilities of the work force to operate in a structured way. For example, some companies believe their core competency may be in milling; thus, every time an option to remove material from a product would allow milling, the milling option is exercised.

The sixth step begins another integrated process design aspect that accommodates an inspection strategy and details the inspection devices. In the concurrent engineering environment of the 21st century, this will also be planned early in the design stages.

The seventh step determines the production tolerances based on form, fit, and function criteria coupled with process capabilities. This also has an aspect of customer acceptance that cannot be underestimated. Results from QFD studies often show that the resultant product design after a QFD analysis will have a much better user interface and much better "look" than was originally detailed by the internal design organization.

The eighth step determines the tools and cutting conditions for the product. Such calculations as the speed of the rotation of the metal removal process, the feed of the material into the process, and the depth of cut of the process are all calculated based on material, surface requirements, and process capabilities.

The ninth step is the overall calculation of the process time (PT). This is *not* the manufacturing cycle time. This is the theoretical "best" by which the processes can be performed and the setup and teardown can be achieved for a batch.

The 10th step involves creating the total process plan on a "process sheet" and creating the numerical control programs for the individual operations. This numerical control aspect of the process design will be developed in Chapter 6.

The above steps are the traditional view of the creation of a process plan. With "time" as the new enterprise metric and with "flexibility" as a critical competitive advantage, we anticipate that the characterization of these form, fit, and function requirements will translate into process requirements that will then be held until the product is going to be produced. When the product will be produced, these process requirements will be extracted from a warehouse of data. They will then be reinterpreted based on availability of pieces of equipment. The flexibility of the equipment will then be exploited, and when we are ready to build the part, we will compare the equipment capabilities with the process requirements and develop flexible equipment assignments. But in this new notion of product-process integration and process requirements versus production capabilities, the need for this *flexible assignment of capacity* will be exploited as we move to modern manufacturing and new manufacturing techniques in a data-driven enterprise as detailed in Chapters 9 and 10. In this section, though, we wish to simply introduce the notion of what is a process plan and how process plans are developed.

A simple example will demonstrate some of the issues Alting portrays. The part is made from a block of aluminum 4 inches by 4 inches by 1 ½ inches. The part is shown in isometric detail in Figure 5–7a and in a three-view sample in Figure 5–7b. As we progress through the steps, the issues will be identified.

***Step 1.*** The interpretation of this design leads the process planner to the realization that several operations must be performed. Also the planner notes that to get to all of the surfaces, the part will probably need to be repositioned during the processing

**FIGURE 5–7a
Process Planning Example**

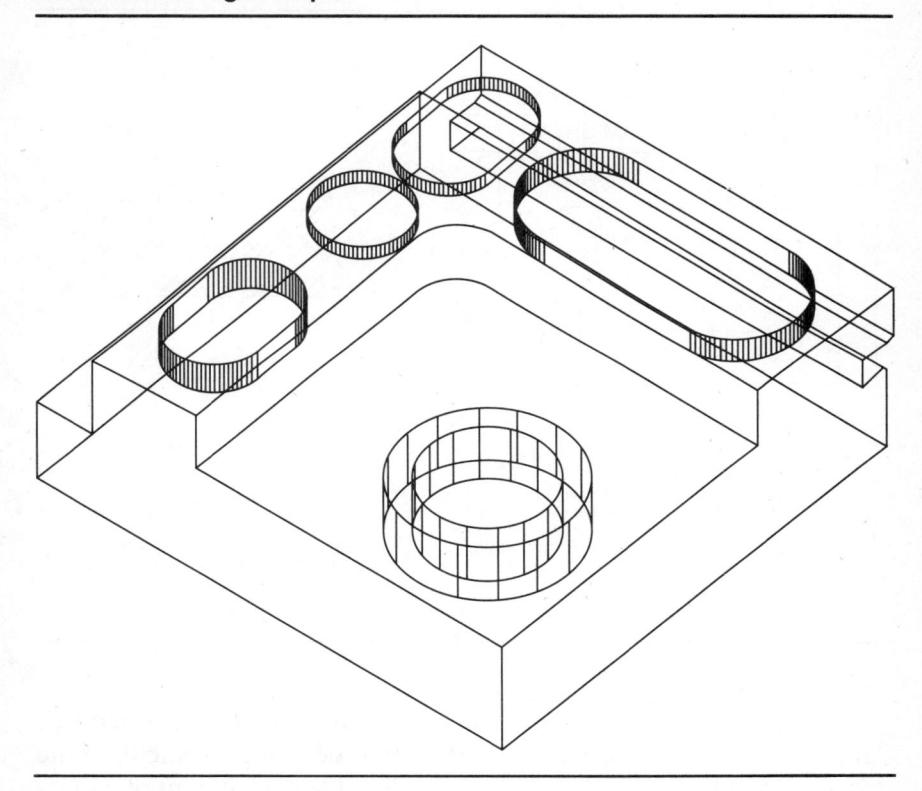

steps. Finally, the interpretation yields that the proper sequence of operations is critical for this part to be made correctly each time.

*Step 2.* We will address only one element of the geometry here for ease of understanding. Notice in a cross-sectional view of section A-A (upper-right-hand corner of Figure 5–7b), it is necessary to mill out three shape elements called "slots." Two of the shapes are oval while the other is a circle. The ovals are milled to different depths. The translation of this geometry to a process requires the analysis of material, shape, depth, and accuracy. An end mill is needed for the shapes and the smoothness of the bottom surface, but the programming of the process must be sensitive to the shape uniqueness.

**FIGURE 5-7b**
**Process Planning Example**

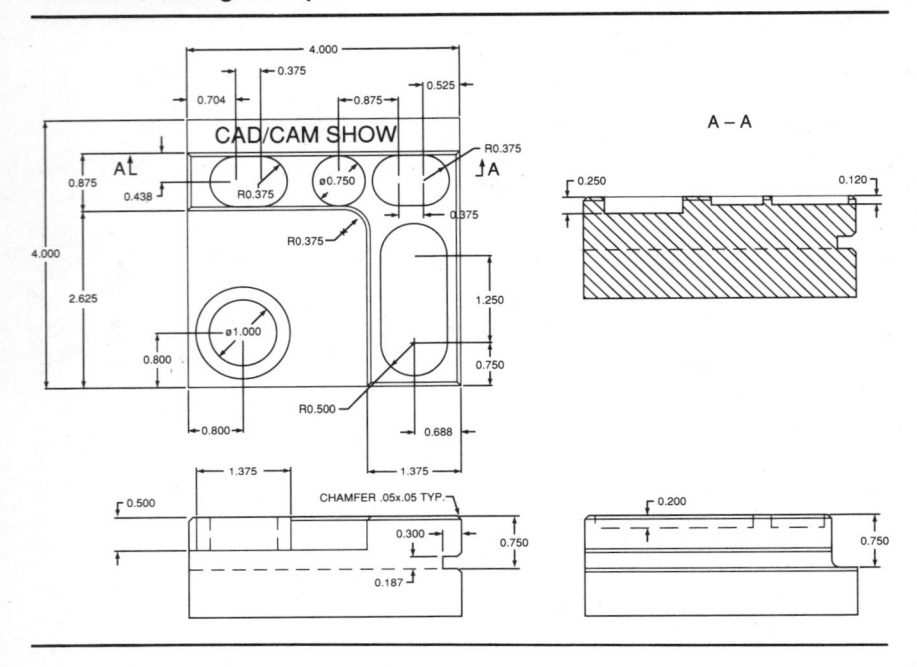

*Step 3.* For our particular shape, a milling machine is needed. Alting recommends that a firm assignment be made to calculate the capacity requirements. In Chapter 9, a flexible assignment of capacity will be investigated in order to assess the real requirements.

*Step 4.* The detailed processing requires that a control surface be established for our milled slots and that the milled slots are referenced off the side dimensions of the block.

*Step 5.* The "best" sequence of operations for this part may be to mill the outside dimensions of the shape first and establish the reference points so that the tooling fixture can be effectively positioned, then insert the end mill with a radius of 0.375 inches. Notice that this tool will be able to make all three slots.

*Step 6.* Inspection must be planned.

***Step 7.*** Tolerances on this print are not stated. The normal way to show a tolerance for this type of process is the desired surface finish on the bottom of the slot and the accuracy of the centers of the slots for the purposes of alignment.

***Step 8.*** We have already selected an end mill with a 0.375-inch radius.

***Step 9.*** The processing time will be calculated using prescribed formulas for the speed of rotation of the end mill, the feed rate for the end mill through the aluminum, and the depth of cut calculation and plunge rate for the end mill into the aluminum block to make the first cut into each slot. This is strictly the theoretical best time.

***Step 10.*** This step captures the steps for the slot creation with the rest of the processes needed for the total creation of this part.

## SECTION 5.7  COMPUTER-AIDED PROCESS PLANNING

Computer-aided process planning (CAPP) was introduced by Mr. Benjamin Niebel in 1965[6] and has slowly evolved. The idea is to use the power of today's advanced computing capabilities to capture the logic that a process planner would use to plan the process outlined in Section 5.6. Unfortunately, many of the 10 steps outlined above are not clear logic patterns but involve detailed knowledge of company strategies and current process capabilities. Because process planning remains a major "art" form in the enterprise environment, even today, use of CAPP techniques will continue to slowly evolve until the process can be standardized and rigorously defined.

The first national attempt to develop a CAPP system was by the nonprofit organization Computer Aided Manufacturing–International (CAM–I). By bringing together industry experts to specify the structure for the system, a "variant" system structure was created that served as the first industry norm.

Today, at least three different notions of process planning methodologies are utilized by manufacturing engineers.[7] The first is called a *variant planner,* the second a *generative planner,* and new methodologies that are called *artificial intelligence-based planners* are evolving.

For the variant planner, the products are typically categorized in terms of generalized families of products. They are characterized by their form, fit, and function into families, and then a process plan format is developed for each family. In this process plan, a series of steps to produce this family of products in a generalized fashion is developed and stored. Once a specific product is defined and identified, the particular process plan part family is called up from the data structure, and it is modified and parameterized based on the specific requirements of that product. Then it is customized to the individual pieces of equipment available in that factory such that a variation, or *variant,* of the product family process plan is developed for each process member of that part family.

With the variant planner, a structure exists for consistency and standardization of process requirements. The translation from a unique product definition in form, fit, and function to the specific process requirement can take on many forms. Surveys have shown that given a detailed, complex product to be planned, 10 process planners will plan the product in 10 different manners. And there is no guarantee that any of these 10 variations are the best given the particular characterization of the product. Hence, the notion that the variant planner can store at least a consensus "best" approach to a process plan for a particular part family is fundamental as we progress to efficient and flexible manufacturing capabilities.

However, if the process requirements can be automatically interpreted from the form, fit, and function and the characterization of the product design, then newer technologies can be exploited to generate automatically the process plan. In this requirement, the product is characterized by its attributes. For example, in our simple metallic part example in Section 5.6, we recognized we needed a machined piece of aluminum and we needed a threaded hole by which we could hold the product to the assembly. These two interpretations, a machining operation and a threading operation, are truly characterized in today's *generative* planning systems by direct interpretation of the solid geometry features into their required metal removal attributes. The attributes are then developed through a decision logic pattern so that they can be translated into specific process requirements directly. Finally, the process plan is developed almost automatically based on the process requirements and the direct translation of these process requirements into equipment specifications.

However, even the generative planners have limitations. They require that a solid geometric model in a digital format representation of the product be present to have the interpretation be automatic from the product definition to the process definition. This solid model quite often is not present, so we are finding that a third approach to process planning is becoming popular. *Artificial intelligence-based* process planning systems are built on decision logic that can be captured in a variety of fashions in many formats and developed in a closed-loop, knowledge-based scenario similar to what Drucker pointed out earlier. If a closed-loop knowledge representation of the process definition can be created and translated back to the product definition in a very timely fashion, then significant time-to-market compression can be achieved. Feedback in a knowledgeable manner through structured decision logic is the essence of the artificial intelligence-based planning systems.

The structure of the variant, generative, and artificial intelligence based systems is based on the recognition that certain similarities exist in the development and capabilities of the product. This was introduced in Section 4.2 of this text.

## SECTION 5.8   PERFORMING THE MAKE-BUY DECISION

Until the notions of core competencies and flexible manufacturing became strategic factors for the enterprise, the decision to make or buy a product was often a "tactical" decision based on capacity and skills criteria. However, if flexible manufacturing capability that can produce any part within a predefined, strategically selected part family can be created, and if the process requirements can be segregated from the equipment assignment, and if strategic core competencies have been defined and are being "grown" within the organization, then the "make-buy" decision has evolved into a strategic decision. The criteria are simple:

1. Does the part fall into a strategic part family within the enterprise's capability?

2. If yes, is capacity available? If yes again, plan to make it!
3. If no to either of the above, then which of our strategically aligned partners has the capability and capacity to provide the product at world-class standards?

Certainly, all decisions are not that simple, but the essence of the decision is strategic and must be addressed as such.

## SECTION 5.9   FIRST ITEM PLANNING WITH CPM, PERT, AND SIMULATION

Budnick, Mojena, and Vollman did a very nice job reviewing the concepts of CPM and PERT from a managerial perspective as early as 1977.[8] CPM and PERT graphically portray all of the activities needed for a complete project. CPM stands for "critical path method," and PERT stands for "performance evaluation and review technique." In this approach, events are defined. An *event* is a specific accomplishment in the overall project plan that can be defined and measured. An *activity* is the work required to complete a specific event. And the network is the linking of these event-activity-event elements. For example:

- "I leave my house" is an event.
- "It takes 30 minutes to drive to work" is an activity.
- "I arrive at work" is an event.

This event-activity-event element is shown graphically in Figure 5–8a. The network can then be constructed to accommodate an entire project as shown in Figure 5–8b. By analyzing and structuring the network of the project, real insight is gained into where *slack* (the difference between the maximum allowed time to perform an activity and the expected duration of the activity) in the project exists for certain activities and which activities fall on the *critical path* (that path through the network where the activities have zero total slack). The critical path is highlighted in Figure 5–8c.

From the critical path network and the project evaluation techniques, key project management details can be developed. These include:

**FIGURE 5-8**
**Critical Path Examples**

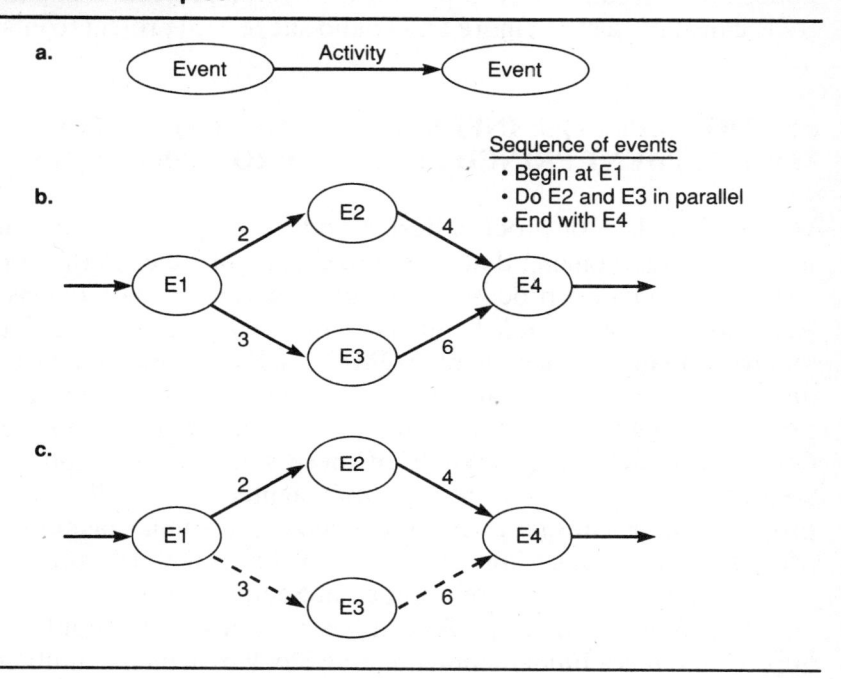

a. Event —Activity→ Event

b. E1 →2→ E2 →4→ E4; E1 →3→ E3 →6→ E4

Sequence of events
- Begin at E1
- Do E2 and E3 in parallel
- End with E4

c. E1 →2→ E2 →4→ E4; E1 --3--> E3 --6--> E4

- Which activities are critical to the timely completion of the project.
- What is the overall resource requirement at any point during the project? Here resource is a critical element of the decision process. Historically, "resource requirements" have been item oriented. But newer views of closed-loop manufacturing resource systems view resources as item needs, skills required, even the cash flow required to achieve success.

Probabilities can be added to the network to accommodate a more realistic project planning scenario, but with the simple network simulation tools of today, this is now done with simulation. The network is digitized based on probabilistic starts, stops, and durations of each activity. By running several Monte Carlo-like comparisons, as developed in Section 4.11, a true distribution and confidence interval can be developed for the overall project. Using these tools can allow a much more timely and successful project to ensue.

## SECTION 5.10 THE INFUSION OF TOTAL QUALITY MANAGEMENT PRINCIPLES INTO PROCESS DESIGN

We have spoken of process design and development as an entity unto itself, recognizing that there are certain fallacies in the ability to totally produce a process design in a rigorous fashion unless we have closed-loop, reliable feedback. This is embraced by the notion that total quality management (TQM) is a major theme that we must understand and incorporate into the process of process design. If a series of steps exists in the product definition through the process design, a model can be created of these steps and the notions of continuous improvement can be implemented. This will allow the product-process design team to recognize in a similar fashion how process design can be improved. The vision, principles, practices, and techniques of TQM are outlined in Section 4.3.

The Malcolm Baldrige Award has recognized that continuous improvement is a fundamental building block by which manufacturing throughout the world can improve. The award is characteristically based on a 1,000-point scale where 300 points are based directly on customer satisfaction. The weight the award places on customer satisfaction is fundamental.

The second notion of TQM is that embraced by Motorola Corporation, the first winner of the Malcolm Baldrige Award. At Motorola, they talk about the notion of the six sigma concept. Six sigma is a corporate commitment to excellence at Motorola. Under six sigma, Motorola develops a model of the process and recognizes where errors can occur. Within this model and analysis, Motorola will allow no more than 3.4 errors per million opportunities for errors.

The major theme for companies like Motorola is that new metrics must be in place. The companies have set very aggressive targets of excellence, developed a model for continuous improvement, and allowed the six sigma notion to serve as an "artifact of iteration." The corporations are recognizing that quality manage-

**FIGURE 5-9**
**Achieving Predictable Behavior**

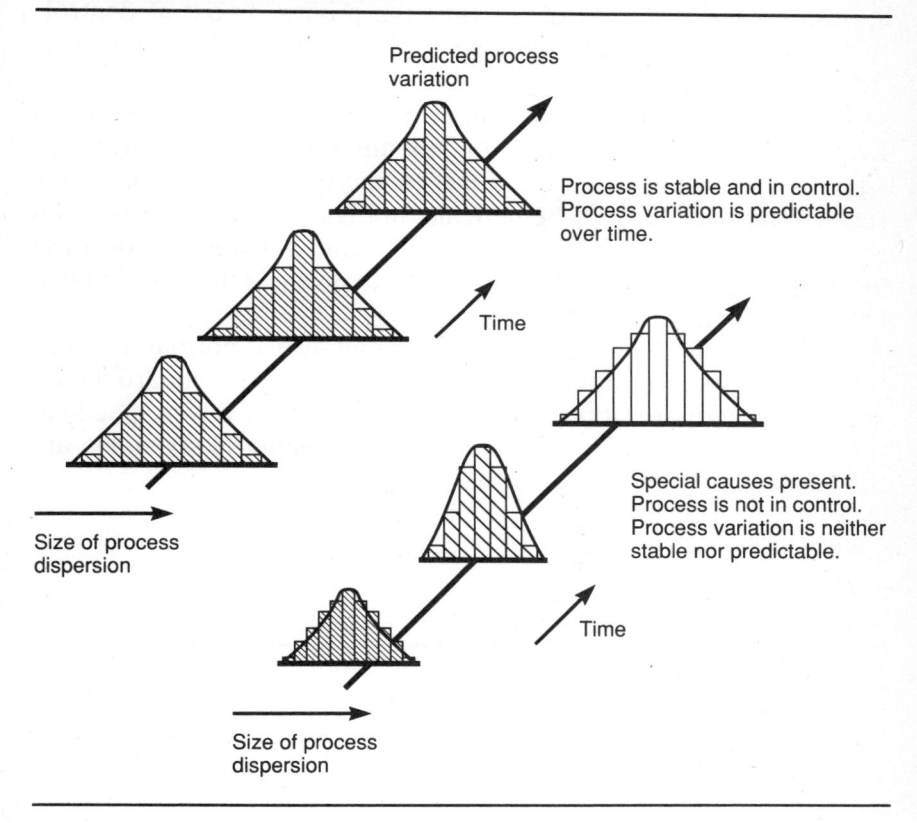

Predicted process variation

Process is stable and in control. Process variation is predictable over time.

Time

Size of process dispersion

Special causes present. Process is not in control. Process variation is neither stable nor predictable.

Time

Size of process dispersion

---

ment is fundamental to their success to serve their customer, and that to do that they must have some metrics in place that can be measured on a continuous basis and then be reestablished to tighter and tighter tolerances so that the company can improve on the way it satisfies customer requirements. The notion of six sigma is that all processes will vary. As seen in Figure 5-9, some processes will have consistent variations that can be modeled and are said to be stable and predictable, while others vary in the sizes of the process dispersions, and these are said to be "not in control."

The objective of six sigma is to get all processes to be fully characterized and, in so doing, to generate stability and predictability that can then be managed and reduced. Based on this variation recognition, if the variation can be managed to the point where less than .00034 percent of the time the process is out of control, then six sigma has been achieved. The calculation for this is as follows:

If the upper specification limit minus the lower specification limit divided by the upper control limit minus the lower control limit equates to a factor of 2, we can then recognize the notion of six sigma. This is depicted schematically in Figure 5–10 where you can see that any process will have an upper and a lower specification limit and all processes will vary in order to produce the part that will fit within these specification limits.

The notion of six sigma is that we can fit *two* normal distributions of process variability (allowing +/− 3 sample standard deviations in each distribution) within these upper and lower specification limits. The formulas for the critical elements of six sigma are the following:

**FIGURE 5–10**
**The Six Sigma Graphical Portrayal**

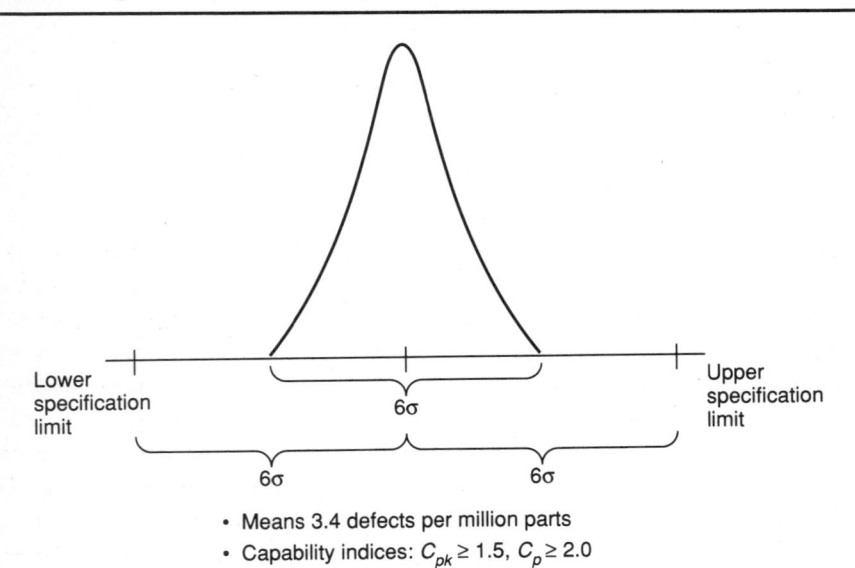

- Means 3.4 defects per million parts
- Capability indices: $C_{pk} \geq 1.5$, $C_p \geq 2.0$

$$C_p = \frac{\text{Upper specification limit} - \text{Lower specification limit}}{\text{Upper control limit} - \text{Lower control limit}}$$

$C_p$ is simply a process capability index. A $C_p$ of 2 represents the target for six sigma. This says the normal variation of each process will occur. But if it can be controlled and/or the specification limits widened, two of the distributions that will exist for a process variation can fit within the specification limits. The definition of the six sigma notion is very simple: "Have no more than 3.4 parts per million errors in anything we do." This 3.4 parts per million is the amount of probability that still remains outside the upper spec limit and the lower spec limit, remembering that the distributions never will touch the x axis. The application of the concept is very, very difficult as we develop process design at six sigma levels.

To achieve the six sigma philosophy, the continuous improvement philosophy, the notion of significant reductions in variabilities of everything we do must be applied to not only the product and the process but also the data and the time available to produce that product. The notion of six sigma applied to data is fundamental to the closed-loop information system requirement for process design in the 21st century. The notion of six sigma on time is not well understood, but it is very clear as we develop the following characterization.

In the sample case study in Chapter 3, the waveguide manufacturing process needed a time for that process. That time is not the process time for a particular item; it is an average process time based on certain characteristics of the process. If we can denote that in Figure 5–11, we can see there is probably a normal distribution of times by which we allow for a basic process to be performed.

We will call that T-bar the mean for that process time from the time we start a process until the time, on average, it will be completed. We will also have variability in our sample size. We will call that simply S-hat, or the variability for the process. Notice that this random distribution will take place in anything that we do.

To focus on continuous improvement, the first thing we must do is focus on process variation reduction. In this particular case, as we look at time for a six sigma application, as we reduce the time variability, we notice two things happening. In the first case, where process variability was large, a process planner and a scheduler

**FIGURE 5–11**
**Applying Six Sigma Notions to Time**

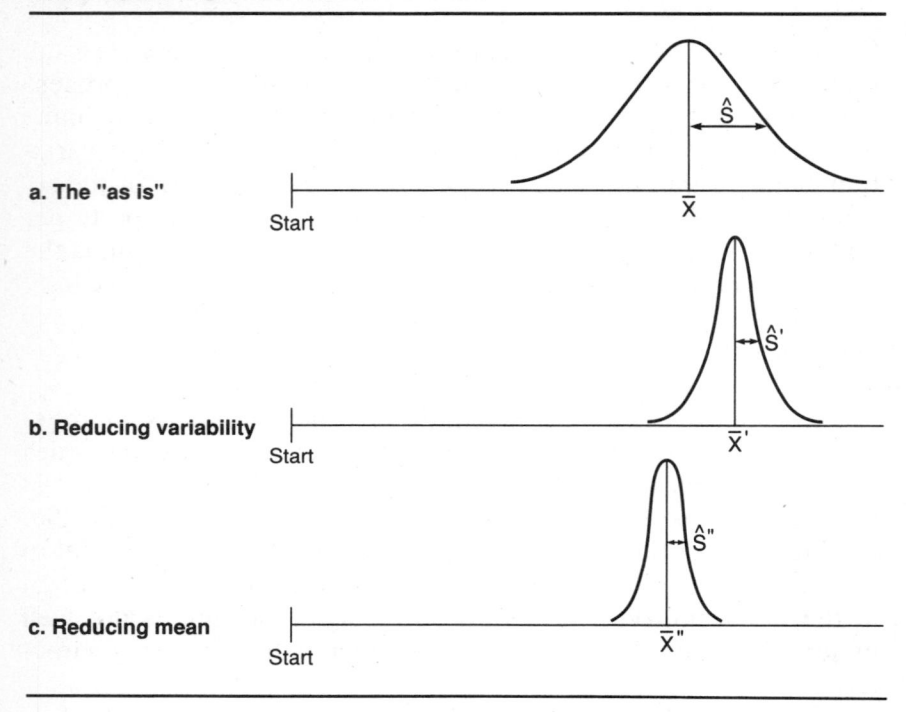

a. The "as is"

b. Reducing variability

c. Reducing mean

would get together and discover they cannot effectively launch that process based on the time on average that it would take to produce that process. In today's environment, we often allow for at least three sigma variation, or three sample standard deviations beyond the mean for the time allowed (in order to accommodate any perturbation).

However, this time between the norm and this variation on the norm results in significant work in process, significant time delays, everything that's bad as recognized in today's world-class manufacturing metrics as shown in Figure 5–11a. Hence, the target of continuous improvement for time is simply to reduce the variation in time it takes to perform any process as shown in Figure 5–11b. The notion here is that if we can reduce the time of variability (even at the expense of allowing the mean times to increase), the time allowed in the process plan will be significantly reduced even though we may not have reduced the average time it takes to perform the process. By reducing the variability on time, significant improvements in work-in-process reduction, characterization, closed-loop feedback, planning functions, and so on can be made. Finally, as we see in Figure 5–11c, once the process is under control, we can begin improving the mean value of the process as well.

The path to continuous improvement can be defined and achieved by addressing the cross-functional team aspects of product and process, and by fully characterizing each of these processes. But this improvement requires more than just plotting process results on $\bar{x}$ charts. In many companies, the statistical process control (SPC) approach successfully highlights the actions of a process and alerts astute operators to make corrections when trends are spotted. However, pro-active control requires pro-active tools like process characterization. With this technique, structured

**FIGURE 5–12**
**Taguchi's Notion of Cost of Quality**

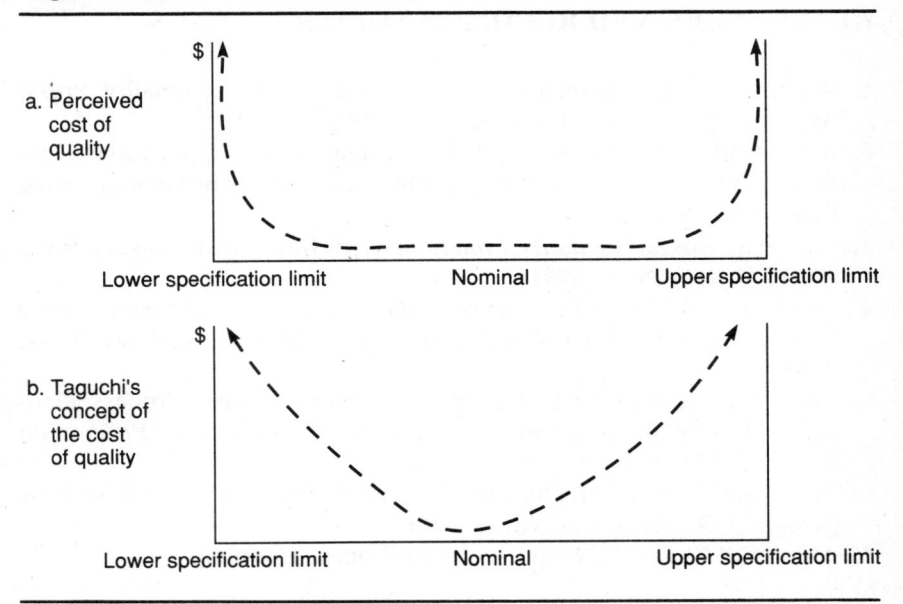

a. Perceived cost of quality

b. Taguchi's concept of the cost of quality

testing using a Design of Experiments (DOE) technique can begin to identify the independent and dependent variables that cause the process variations. This more sophisticated statistical approach is based on the hypothesis that any variation from the nominal, desired value actually does cost money. Taguchi[9] initiated this concept when he introduced the "Cost of Quality" and his "Loss Function." Figure 5-12a shows how the historic perception of quality allowed the process to vary anywhere between the upper and lower specification limits with *no* detriment to the fit of the products, hence no added cost for the quality. But Taguchi showed that *any* variation from the nominal design optimum value is detrimental to the fit of the products, hence adding costs. As a matter of fact, these costs escalate up such a steep slope that the cost curve is really a parabola (Figure 5-12b). Staying close to nominal is the most cost-effective total system cost that can be achieved, Taguchi argued. This is clearly consistent with Drucker's opening comments for the factory of 1999.

## REFERENCES AND RECOMMENDED READINGS

1. Robert C. Camp, *Benchmarking* (Milwaukee, Wis.: Quality Press, American Society for Quality Control, 1989).
2. A. J. Roth, Jr., "Productivity by Design—FMS Applications That Work," Dearborn, Mich.: FMS Conference, Society of Manufacturing Engineers, 1986.
3. James L. Bossert, *Quality Function Deployment* (Milwaukee, Wis.: ASQC Quality Press, 1991).
4. Dan L. Shunk, *Rapid Prototyping Facilities in the U.S. Manufacturing Research Community* (Washington, D.C.: National Academy Press, 1990).
5. Leo Alting and Hongchao Zhang, "Computer-Aided Process Planning—A State of the Art Survey," *International Journal of Production Research* 27 (1989).
6. Benjamin Niebel, "Mechanized Process Selection for Planning New Designs," ASME Paper No. 737, 1965.
7. Alting and Zhang, "Computer-Aided Process Planning."

8. Frank S. Budnick, Richard Mojena, and Thomas Vollman, *Principles of Operations Research for Management* (Homewood, Ill.: Richard D. Irwin, Inc., 1977).
9. Genichi Taguchi, Elsayed Elsayed, and Thomas Hsiang, *Quality Engineering in Production Systems* (New York, N.Y.: McGraw-Hill, 1989).

# CHAPTER 6
# CONTROLLING THE PROCESS

Chapters 4 and 5 developed the prerequisites for integrated process design and development and the ability to create the process plan. You now have a fairly good understanding of what is required to plan the process. However, you must still understand how to convert that process plan into an actual part.

This chapter focuses on the production of a metallic part, on how "control" of the process and of the assets can be achieved. The concepts presented in this chapter are directly applicable to the notions of:

1. Composites lay-up where a numerical control tape is also developed using the process plan notation to create a composite product through a numerical process.
2. Semiconductor processing where precise alignment of one mask layer on top of another requires sophisticated numerical control.
3. Electronics processing where numerically controlled insertion equipment allows one to first drill the printed circuit boards properly and then locate, position, and insert electronic components onto a printed circuit board.

## SECTION 6.1   INTRODUCTION TO CONTROL

Engineering is concerned with understanding and controlling the materials and forces of nature for the benefit of mankind. The control system engineer is concerned with understanding and controlling a segment of his or her environment, often called a *system,* to provide a useful economic product for society. The twin goals of understanding and control are complementary because, to control more effectively, the system under control must be understood and modeled. Furthermore, control engineering often must consider the control of poorly understood systems such as chemical process systems. The present challenge to control engineers is the modeling

118  Chapter 6

and control of modern, complex, interrelated systems such as traffic-control systems, chemical processes, and economic regulation systems. However, simultaneously, the fortunate engineer has the opportunity to control many very useful and interesting industrial automation systems. Perhaps the most characteristic quality of control engineering is the opportunity to control machines and industrial and economic processes for the benefit of society.

Control engineering is based on the foundations of feedback theory and linear system analysis, and it integrates the concepts of network theory and communication theory. Therefore, control engineering is not limited to any engineering discipline but is equally applicable for aeronautical, chemical, mechanical, civil, and electrical engineering. For example, a control system often includes electrical, mechanical, and chemical components. Furthermore, as the understanding of the dynamics of business, social, and political systems increases, the ability to control these systems will increase also.

**FIGURE 6–1**
**Focus on the Control Functions**

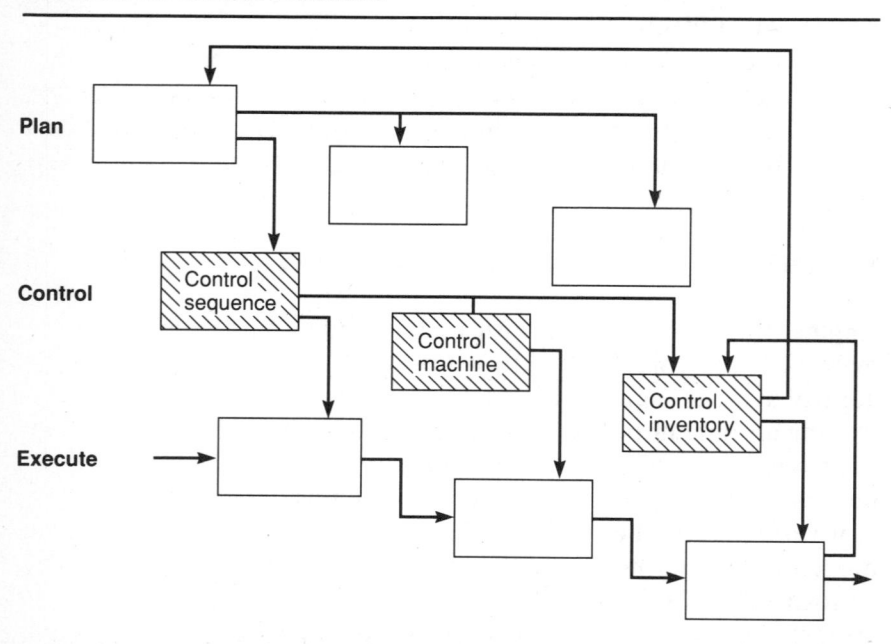

Mr. Richard Dorf states, "A *control system* is an interconnection of components forming a system configuration which will provide a desired system response."[1] The basis for analysis of a system is the foundation provided by linear system theory that assumes cause-effect relationship for the components of a system. Therefore, a component or process to be controlled can be represented by a block as shown in Figure 6–2a.

The input-output relation represents the cause-and-effect relationship in the process, which in turn represents a processing of the input signal to provide an output signal variable, often with a power amplification. An *open-loop* control system utilizes a controller or control actuator to obtain the desired response, as shown in Figure 6–2b.

In contrast to an open-loop control system, a closed-loop control system utilizes an additional measure of the actual output to compare the visual output with the desired output response. A simple *closed-loop feedback control system* is shown in Figure 6–3a. A feedback control system is a control system that tends to maintain a prescribed relationship of one system variable to another by comparing functions of these variables and using the difference as a means of control. The feedback concept has been the foundation for control system analysis and design.

Because of the increasing complexity of the system under control and the interest in achieving optimum performance, the importance of control system engineering has grown in this decade. Furthermore, as the systems become more complex, the inter-

**FIGURE 6–2**
**Control Block Diagram and Open-Loop Control**

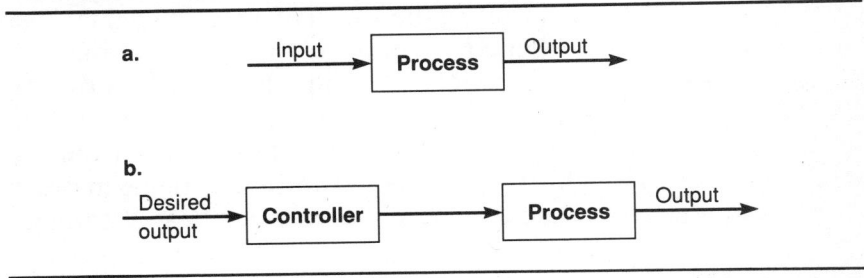

**FIGURE 6-3**
**Closed-Loop Feedback and Multivariate Control**

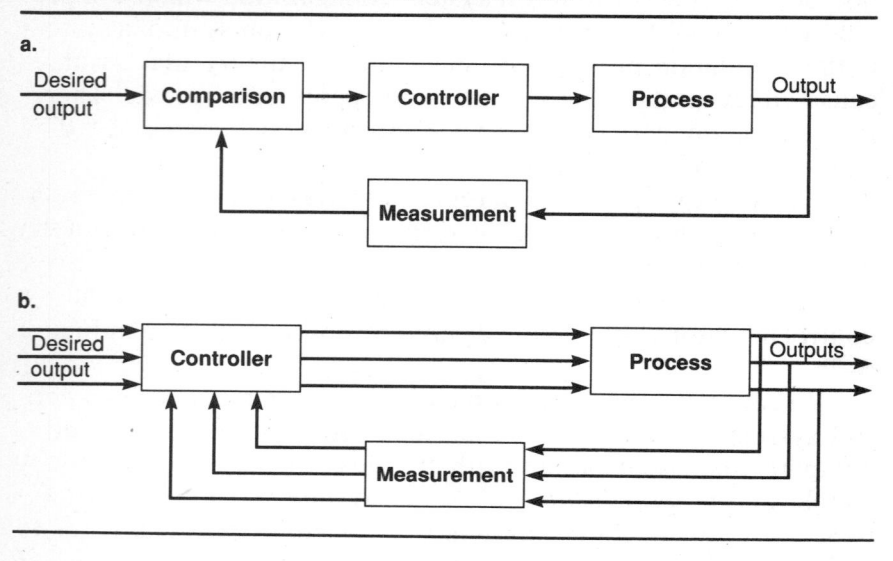

a.

Desired output → Comparison → Controller → Process → Output
                     ↑                                    ↓
                     └──────────── Measurement ←──────────┘

b.

Desired output → Controller ⇒ Process ⇒ Outputs
                     ↑↑↑                    ↓
                     └──── Measurement ←────┘

relationship of many controlled variables must be considered in the control scheme. A block diagram depicting a *multivariable control system* is shown in Figure 6–3b.

## SECTION 6.2  HISTORY OF AUTOMATIC CONTROL

The use of feedback to control a system has had a fascinating history. The first automatic feedback controller used in an industrial process is generally agreed to be James Watt's flyball governor developed in 1769 for controlling the speed of a steam engine. The all-mechanical device, Figure 6–4, measured the speed of the output shaft and utilized the movement of the flyball to control the valve and therefore the amount of steam entering the engine. As the speed increased, the ball weights rose and moved away from the shaft axis. The flyweights required power from the engine in order to turn and therefore made the speed measurement less accurate.

**FIGURE 6–4**
The Flyball Governor

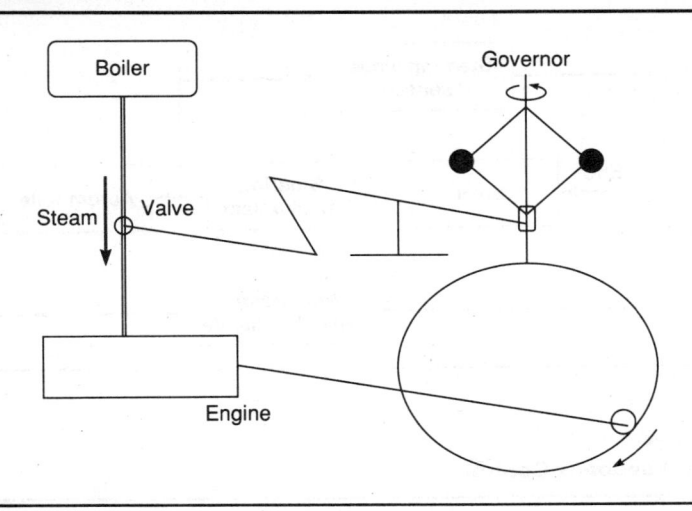

## SECTION 6.3 EXAMPLES OF MODERN CONTROL SYSTEMS

Feedback control is a fundamental fact of modern industry and society. Driving an automobile is a pleasant task when the auto responds rapidly to the driver's commands. Many cars have power steering and brakes that utilize hydraulic amplifiers on the force applied to the brakes or the steering wheel. Figure 6–5 shows a simple block diagram of an automobile steering control system. The desired course is compared with a measurement of the actual course to generate a measure of the error. This measurement is obtained by visual and tactile (body movement) feedback. Additional feedback comes from the feel of the steering wheel by the hand (sensor). This feedback system is a familiar version of the steering control system in an ocean liner or the flight controls in a large airplane.

The system shown in Figure 6–6 is a *negative feedback* control system because the output is subtracted from the input and the difference is used as the input signal to the power amplifier.

Finally, it is possible to model the feedback processes preva-

**FIGURE 6–5**
**Automobile Steering Control**

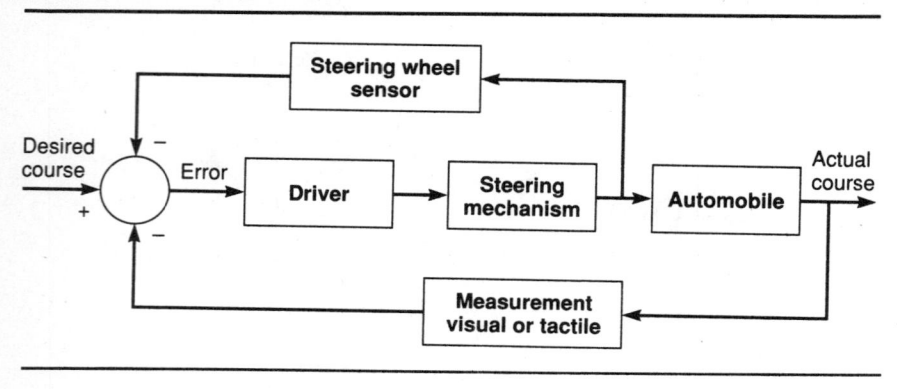

**FIGURE 6–6**
**Negative Feedback Control**

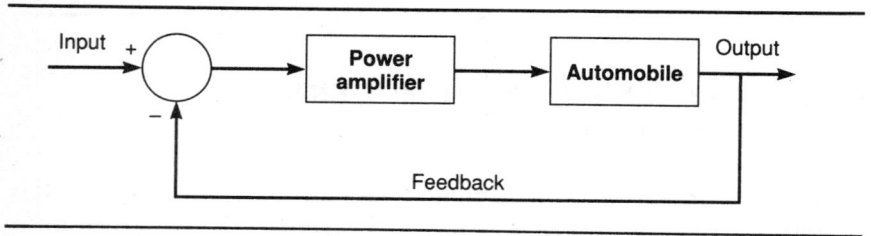

lent in the social, economic, and political spheres. Society is comprised of many feedback systems and regulatory bodies such as the Interstate Commerce Commission and the Federal Reserve Board that exert the necessary forces on society to maintain a desired output. A simple lumped model of the national income feedback control system is shown in Figure 6–7.

This type of model helps the analyst to understand the effects of government control and the dynamic effects of government spending. Many other loops also exist, since, theoretically, government spending cannot exceed the tax collected without a deficit, which is a control loop containing the Internal Revenue Service and the Congress. This type of political or social feedback model, while usually nonrigorous, does impart information and understanding.

**FIGURE 6-7**
**An Economic Closed-Loop System**

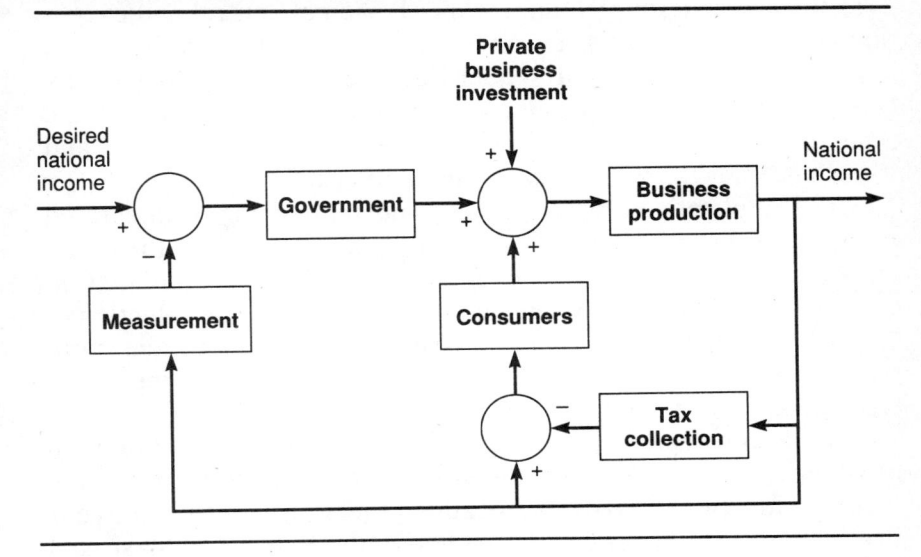

The potential future application of feedback control systems and models appears to be unlimited. The theory and practice of modern control systems have a bright and important future and justify the study of modern automatic control system theory and application.

## SECTION 6.4  CONTROLLING THE FLOW OF PRODUCT THROUGH THE FACTORY

Both MRP I and MRP II load the factory based on a plan and launch the orders often in batches. Several popular modern MRP II systems have scheduling modules. However, in the 21st century factory, a schedule removed from the actual factory status by one day to one week shall be of limited value. Actual factory scheduling and tracking is better done with on-line interactive systems rather than the batch operations of many systems today. And most certainly, the *assignment* of the correct, available machine and the *sequencing* of jobs onto these machines are control functions done when the actual operations need to be performed. In many just-in-time environments, the MRP system is not used at all for the factory. MRP is used for longer-term planning, while the actual control and execution is conducted much closer to the factory floor with on-line systems. Assignment and sequencing are classically defined by David Bedworth and James Bailey as: "the assignment of machine tools to accomplish each operation on each scheduled part," and "in what order the waiting parts should be run."[2]

The implications of assignment and sequencing to integrated process design and development are significant. By using the group technology methodology and SMED methodology outlined in Chapter 4 for design of tooling and fixturing and for the selection of flexible manufacturing systems, assignment now can have more freedom and sequencing can now support the downsizing of lot sizes to achieve flexibility.

Modern research is viewing this factory control function with great interest because it is the first step in the transition from planning functions to true, value-added functions. The objective is to develop tools that reduce the lead times through task analysis and activity relationships in an integrated process development environment. By overlapping tasks and by utilizing duplicate capabilities, time compression can be achieved. Once a model of the sequence of operations needed to create a product is defined, task analysis can begin by viewing each task as a set of subtasks and determining which of these subtasks must be performed serially and which are candidates for parallel efforts. The advantages of this are:

1. Lead time and production costs can be minimized.
2. Through the decomposition, groupings can be much more logical and efficient.
3. Human networking now receives the mental benefits from an efficiently operating environment.

By incorporating task analysis with the technique of numerical control, developed in Section 6.5, the factory is now in a position to control the flow efficiently and effectively while achieving a significant level of flexibility.

## SECTION 6.5  THE HISTORY OF NUMERICAL CONTROL

The historical background for numerical control is quite interesting. It began in the late 1940s with the U.S. Air Force through the Air Force Manufacturing Technology Division at Wright-Patterson Air Force Base. The division sponsored a series of research projects that ultimately led to some work being conducted at Massachusetts Institute of Technology with the active participation of Dr. Joseph Harrington, Jr., and at the Illinois Institute of Technology Research Institute with the active participation of George Putnam. The teams at MIT and at IITRI developed a radical approach to the ability to produce machine parts.

In this approach, they recognized the possibilities of using a computerized system. They also recognized that in 1948 to 1953, computers were bulky and crude. They noticed that if they had the ability to control the machine tool through a process using digital definition of programs, they would have the ability to develop much more consistent and reliable processes and allow the machine tool operator to concentrate on planning and setting up operations without having to constantly worry about every axis on a multi-axis machine tool turning simultaneously. Their developments were demonstrated in a prototyping activity in 1952. However, development of the machine control mechanism created a need for a language to support this numerical control. They reasoned that if each potential user developed his or her own communication methodology, the industry would still not have the standardized approach necessary to establish a national standard, and ultimately an international standard, of numerical control concepts in the factory.

Hence, a second concept was developed called the "APT" language. APT stands for "advanced, or automated, programming tool." APT allowed the planners to translate the geometry of the product and the process plan of the product into digital definition in a consistent fashion such that machine tool controllers could be developed to translate this into the actual operation of the machine tool.

But a third element was still necessary. To jump start this instantiation of advanced technology into the machine tool industry, the U.S. Air Force recognized that a "bulk buy" of machine tools was necessary to create a big enough market so developers could amortize the development cost effectively and sufficient NC machine tools could be produced and distributed to truly test the concept. Hence, a third major element of the history of numerical control was the U.S. Air Force Manufacturing Technology Program's bulk buy of machine tools that were inserted into the aerospace industries in the mid-1950s. This presented enough marketplace for the numerical control machine tool industry to get started and is a significant factor in the advancement of manufacturing automation and integration today.

In the following sections, the notion and the understanding of numerical control is introduced, not necessarily to make the reader an expert in being able to create an APT program, but to aid in the understanding of why the process was developed and also what the process is and how it relates to integrated product-process design and development methodology. We also hope to introduce the notion of how one can actually take a geometric definition of a part and the process plan and, ultimately, create machine controls that can produce that part.

Finally, we hope to introduce the notion of closed-loop control that uses advanced concepts of sensing and other technologies to ensure that the process being performed is the process that was prescribed.

## SECTION 6.6  MACHINE CONTROL CONCEPTS

The significant changes occurring in the workplace today are pushing the world toward closed-loop, integrated process control. The need for time management à la Drucker and the ability to achieve significant flexibility in the production of all of our parts, to such an extent that we have the ability to produce an economic order quantity of one in a cost-effective manner; the ability to produce total quality in our products for a consistency of the process at the six sigma level of qualification; and the ability to understand how tolerancing comes into play in our total quality scenarios are very important. These quality characteristics require significant rigor in the overall control of the process and hence the need for a digitized manner to control the machine.

Finally, as we discuss in this chapter the notion of integrability, it is fundamental to recognize that machine control concepts for the 21st century must be integrated fully throughout the entire world. To be integrable, machine control concepts must be standardized and they must have the same characteristics in terms of definition of products, languages by which processes will be defined, and overall machine control concepts. To achieve this, the machine control of the process is fundamental for our overall success.

As Mr. Mikell Groover and Mr. Emory Zimmers highlight in their CAD/CAM book of 1984, five distinct steps are necessary to utilize numerical control in manufacturing.[3] These are:

1. The ability to create the process plan. We demonstrated in Chapter 5 how this process plan can be developed, and we will build on that development process in this chapter.
2. The ability to translate the process plan into a digital definition of a product that can be utilized by a machine tool. This we will call numerical control, and extensions will be developed later.
3. The translation of the numerical control technique into the ability to create a data file, whether it be on Mylar tape or whether it be in a digital format, such that we can translate the part program into a certain file by which the machine tool can then accept the part program and verify that the program makes sense.
4. The ability to translate specific offsets and the ability to translate specific machine tool characteristics into the control language so that the generic data file is now made specific for the specific machine tool that's now in existence.
5. The utilization in production by which setup, tooling and fixturing are all considered for the machine tool operator to ultimately accept and be able to manage the integrated product and process design and development scenario.

Let us look at each of these in a bit more detail.

First, we must create the process plan, and as we saw in Chapter 5, this can take on many forms. In a variant process planning system, we look at modifying a standard process plan that's in a data file to add specific characteristics, geometries, and so on to the process in order to specify exactly how to produce that process. In a generative mode, a geometry file is translated directly through knowledge-based systems into the overall process to create the part. However, in each of these, each of the steps, such as mill, or drill, or turn, must be specifically detailed in the overall control of the process. Taking each of the specific steps of the process plan and translating it into the specific detail by which the process will be controlled is the next level of detail in this text. Taking each step of the process and translating each of those steps in a consistent manner creates the ability to support the flexibility, consistency, and integrability necessary to achieve this integrated process design and development scenario.

Step two is the translation of the geometries, tolerances, surface finishes, and so on into a rigorously specified syntax that creates a part program. This is normally done using a language like APT and contains four types of statements:

- Geometry statements, such as "P1 = Point/5.0,4.0,0.0."
- Motion statements, such as "GoTo/P1."
- Postprocessor statements, such as "COOLNT."
- Auxiliary statements, such as "CUTTER."

Step three processes the program into machine readable, compiled code for use by the machine controller.

Step four adds specific notations concerning the parameters for the machine tools to be used, including cutter offsets, machine tool characteristics including the NC processor at the tool itself, and the location of each of the cutting tools.

Finally, step five is the actual running of the program on the tool after the setup has been performed, after the cutting tools have been loaded, and so forth. By loading the tape or downloading the program from a supervisory computer, each part is then produced as a unique entity supporting the notion of an economic order quantity of one.

To achieve this integrability, we must have consistency in the process and the routings. In the material management world, these will be called the *routers,* and they will be used in the MRP or scheduling systems, the manufacturing resource planning and the material requirements planning systems. These routings will be used to define what materials are going to be required and, ultimately, to determine the capacity requirements. However, in the engineering world, these will be called the *process plans*. They are the same, although the process plans will have much greater detail than the routers.

Every MRP system is driven off these process plans, but every MRP system accepts only one process plan that is prime and located typically in some sort of a computerized data structure. It does not matter at this time how to achieve this flexibility, but as we have identified and will develop in Chapters 9 and 10 of this text, we perceive significant changes in the ability to store this process plan and utilize it throughout the world by having generic structures available.

## SECTION 6.7   PRINCIPLES OF OPERATION

While many different machines are numerically controlled, the principles of operation have much in common. In fact, relatively few different control systems exist. Mr. Harry Moore and Mr. Donald Kibbey point out that it should be kept in mind that the subject is numerical *control* and that the systems being controlled are often conventional machines that might otherwise be controlled manually.[4] Even when new machine tools have been built with numerical control, they are still machines that do conventional machining operations and use the same cutting tools and motions that have been discussed in previous chapters.

Several different media have been used for input data, including punched tape, punched cards, magnetic tapes, and direct input from computers. The progress in standardization, which usually accelerates the rate of adoption of new systems, has been much more rapid than in past machine tool developments. Even with the wide diversity of numerically controlled machines, the NC revolution began with the general acceptance of a Mylar punched tape, a one-inch wide, eight-channel tape used as the standard input medium. Figure 6–8 shows a section of such a tape. The presence or absence of a hole in a particular location is referred to as a *bit* of information. A row of holes, or possible hole locations, across the

**FIGURE 6–8**
**Sample NC Tape**

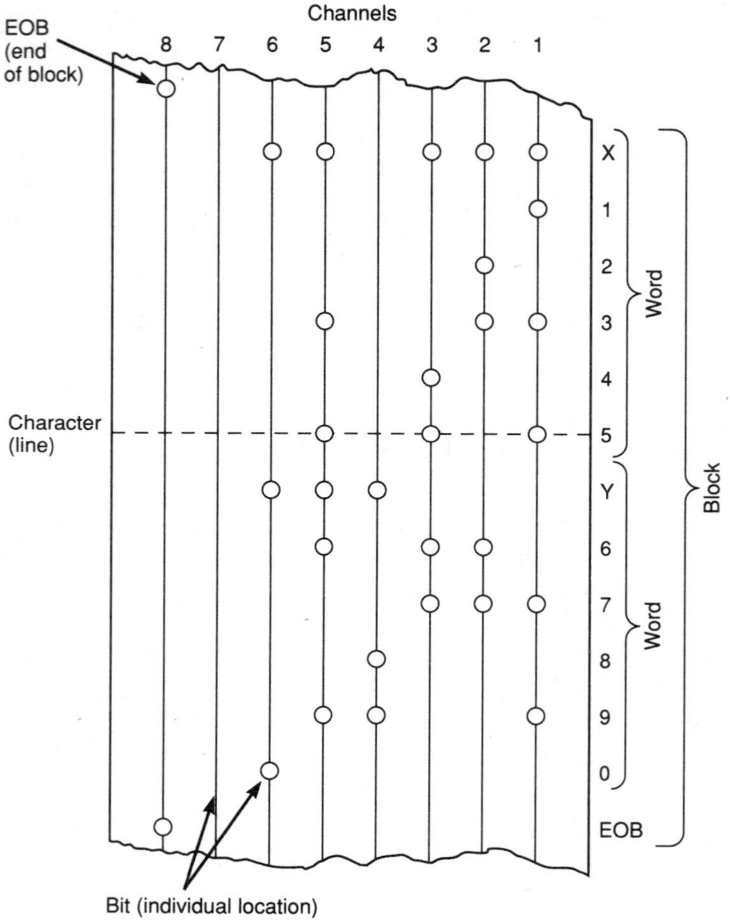
Channels
EOB
(end
of block)
8 7 6 5 4 3 2 1
X
1
2
3
4
5
Y
6
7
8
9
0
EOB
Word
Word
Block
Character
(line)
Bit (individual location)

width of the tape constitutes a *character,* which is normally a number from zero to nine or one of the letters of the alphabet.

A standard binary code is used to form the numbers. A hole in the first channel has the value one, a hole in the second channel the value two, a hole in the third channel the value four, and a hole in the fourth channel the value eight. Tape readers can be designed to examine each character electrically to check whether an odd or even number of holes is read. Each character is made to have an odd number of holes on the tape so that if the reader reads an even number of holes, the reading error can be appropriately indicated or the machine stopped automatically. To make characters have an odd number of holes when the binary code calls for an even number, a hole is punched in the fifth channel, as is done for the letters X and Y, and the numbers 3, 5, and 6.

Letters are made of combinations of holes in the first, second, third, fourth, sixth, and seventh channels. A hole in the eighth channel is used only to indicate the end of one complete block of words, usually corresponding to one machine movement. Not shown in the figure is a row of smaller holes between the third and fourth channels that is used for feeding the tape.

Regardless of their complexity, most numerically controlled machines have somewhat similar control and drive systems. Most contain a number of independent elements each controlled by one word in a block of taped information. These elements are of two distinct types. Auxiliary functions such as coolant supply, spindle rotation, spindle speed, and tool selection have a limited number of discrete values possible (coolant on or off, for example) and are controlled by open-loop systems such as shown in Figure 6–9a. Some actions associated with driving an automobile may be compared to these open-loop auxiliary functions. These include unlock the door, turn the switch key on, place the transmission in second gear, turn headlights on, or blow the horn. For each of these actions, a fixed set of motions must be performed, and there is no other control. If the driver operates the light switch and the lights do not come on, the driver may operate as an interlock and stop the machine, but so far as turning on the lights is concerned, the driver has no corrective action to take other than repair.

The second type of numerically controlled function is much more complex. For machine tools, this is concerned with the variable motions of machine slides that position the piece for drilling or other point-located operations, feed the slide for milling operations, feed the carriage or cross slide on a lathe, or control similar motions on other machines. Some machines have been built with open-loop systems for control of these motions, but the majority used closed-loop systems such as shown in Figure 6–9b.

**FIGURE 6–9**
**Open-Loop and Closed-Loop Machine Control**

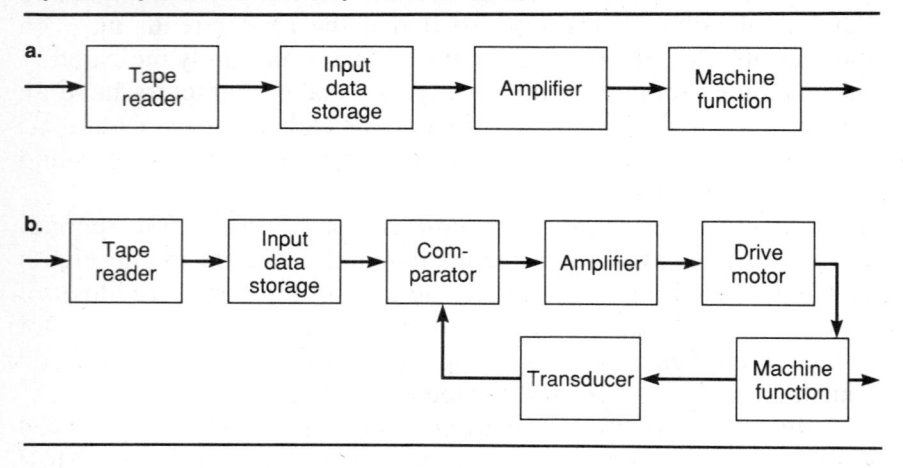

a. Tape reader → Input data storage → Amplifier → Machine function →

b. Tape reader → Input data storage → Comparator → Amplifier → Drive motor → Machine function →
                                          ↑                                          ↓
                                          └──────── Transducer ←─────────────────────┘

Inherent in each closed system is a *transducer,* which is connected to and driven by the machine component (slide) under control. Transducers are of several types, but most operate by generating an electrical pulse each time the slide moves some given increment of distance, the size of which is normally 0.001 inch or less, with 0.0002 inch being most common. The polarity of the pulses is reversed when the direction of slide movement is reversed. The comparator in the system contains a counter that adds or subtracts pulses, so that it is constantly supplied with information as to the location of the slide. The actual location can be continuously compared with the desired location (tape information stored in the memory of the comparator) and corrective action taken until coincidence occurs.

A single machine motion consists of the following sequence. The tape reader feeds a desired new location for the slide through the data storage section to the comparator. Since this new location does not correspond to the present position of the slide, an *error* signal is generated by the comparator. The error signal is fed through the amplifier to the drive motor, and the machine slide travels in such a direction as to reduce the difference between the slide position and the desired new slide position. When the value in the comparator from the machine slide corresponds to the desired new value, the error signal becomes zero and the drive stops. The system is made more complex by the necessity of traveling at reasonably high rates of speed to the new position and stopping accurately at the position, preferably without overshooting.

## SECTION 6.8  TYPES OF CONTROL SYSTEMS

As in conventional machine tools, the total relative motion between the cutter and the work on a numerically controlled machine is made up of a number of simple motions, either straight lines or rotation about axes. On an engine lathe, control of the cross slide on the carriage and the carriage along the bed is sufficient for all standard operations. Such a machine would be known as a *two axis* machine. Milling machines may have as many as five axes controlled: linear motion along three orthogonal axes and rotation about two axes.

For dimensioning and programming purposes, it is common to refer to the linear axes as X, Y, and Z, with X and Y in the horizontal plane and Z in a vertical direction. Two different dimensioning systems are used. In the *absolute* system, dimensions are stated with reference to fixed baselines, one for each axis. The baselines normally have the value zero. Most machines using an absolute dimensioning system are equipped with *floating zero,* that is, the control system may be set so that the value in the comparator from the transducer may be made zero at any point within the travel limits of the table. With an *incremental* control system, the dimensions for each point must be stated with reference to the last point programmed. In effect, the baselines shift each time a new point is reached to correspond to that point.

Regardless of the number of axes, numerically controlled machines are classified either as *point to point* or as *continuous path,* depending on the way in which motions involving two or more axes are made. In Figure 6–10, $X_1Y_1$ and $X_2Y_2$ represent two points on a piece between which a motion must be made. These points could represent hole centers for drilling, points along the contour of a part to be turned, or points on a path to be followed by a milling cutter. The X direction could represent the motion of the cross slide and the Y direction the motion of the carriage on a lathe. Since the line

**FIGURE 6–10**
**Defining the NC Dimensions**

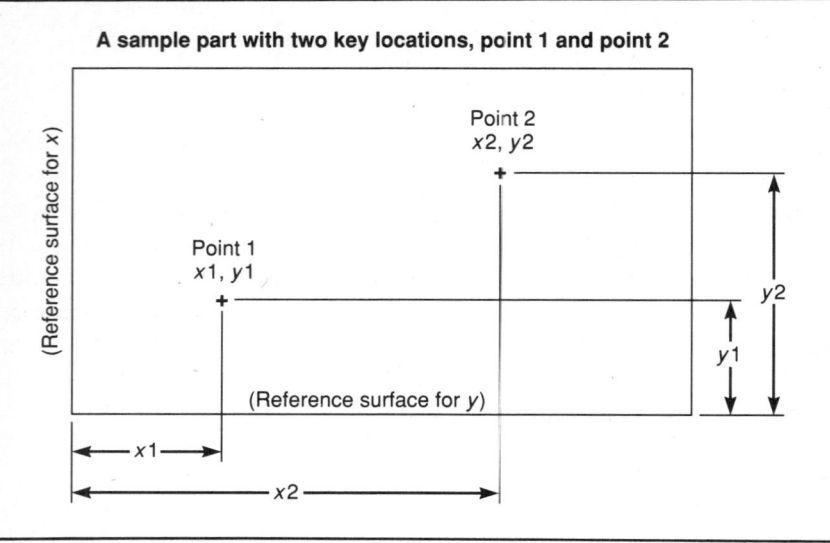

A sample part with two key locations, point 1 and point 2

(Reference surface for x)

Point 2
x2, y2

Point 1
x1, y1

(Reference surface for y)

x1

x2

y1

y2

between the two points does not correspond exactly to a single motion of either slide, both slides must move for the cutter to travel from point one to point two.

## SECTION 6.9  THE CREATION OF THE GEOMETRY PROFILE FOR THE PROCESS

The second step in the overall development and creation of this process plan is the ability to translate from the geometry profile and the process plan into a part program, which we will now call a *numerical control program*. In this, we take the geometry definition of the product and identify what processes must be performed. Then the planner translates the process requirements into machine readable structuring using a programming language called APT or advanced APT. APT stands for advanced programming technique or automated programming tool. APT is the communication language for the numerical control process plan. APT is the ability to translate the geometry via a coordinate system into "go-to" statements that, in essence, dictate where the machine should move, how it should move, in what fashion it should move, what processes should be turned on, what processes should be turned off, and so on. It is not necessary in this text to define how APT operates.

What we are attempting to do here is fundamentally important. We are attempting to develop a common cutter location file, or CL file, by which all machine tools can interpret consistently the geometry that needs to be developed based on a certain process requirement. For this to be successful, certain roles must be played out consistently. For example, an individual or team of individuals responsible for this translation must define the geometry of the workplace and also the operation sequence and the tool path such that a consistent program definition can be developed using this language.

The second is the role of the machine tool controller. We are now ready for the translation of the process plan and the overall part geometry into the data file, which we will call the CL, or cutter location, data file. We must take the promptings from the part programmers tasks and translate them through a rigorous analytical methodology that does arithmetic calculations to the input data. The part program translator outputs the numerical control language that is read by the machine controller.

We must accommodate the individual differences of the machine tools we wish to control. In step four, we translate the specific machine characteristics from this generic numerical control tape language. This is called tool offsets. Ultimately, post-processing of data will develop a language that allows the specific machine tool with its specific controller, specific configuration, specific size of process envelope, and specific configuration of tools and fixtures to accommodate the geometry and the processing of the part that have been planned using our APT and our CL data file. We have been successful in the past at utilizing standardized methodologies to define what this overall system is.

Finally, we will look at the overall production and recognize that tooling and fixturing, setups, and so on are still the responsibility of the numerical control machine tool operator. It is this operator's responsibility to ensure that the parts are set up on the beds in a correct fashion and that the process is being performed to its ultimate extent. In the past, the machine tool operator not only set up the machines but also turned all spindles and performed the process himself or herself. This fundamental shift to a monitoring and control role for the operator has forced advanced requirements on these operators to understand not only the operation of the machine tool but also the arithmetic operations of the control language and the computer operations of the controller on that machine tool.

This process requires significant integration from the design of the product, to the planning of the process, to the control of the process, to the ultimate execution of the process at a specific, predefined machine tool with specific setup and fixturing. This allows for the integrability and also the transportability of these programs throughout the factory floor as well as across the country and throughout the world. But the ultimate objective here is to provide consistent, rigorous process definition data so that the flexibility of an economic order quantity of one, the efficiencies, and the effectiveness can be managed in one singular entity.

## SECTION 6.10 ADVANCED NUMERICAL CONTROL DEVELOPMENTS

The advent of numerical control in the late 1940s to mid-1950s has dramatically affected the overall manufacturing base of the world. The ability to produce consistent parts with specified processes in a consistent fashion has revolutionized the requirements and the abilities to provide six sigma process quality on all of our machine tools and to hold very, very tight tolerances to produce very, very precise pieces of equipment that have allowed us to advance the overall state of the art.

Now, recent advancements in the integration of the numerical control language with interactive graphics and advanced technologies for integrability for computer numerical control (CNC) and direct numerical control (DNC) have added another level of detail and sophistication to controlling the process.

To translate and create the cutter location data file, we must translate the geometry into specific steps by which the part programmer translates the geometry of the product. This translation lends itself to errors. The industry has recognized the distinct need to integrate the geometry from the computer-aided design (CAD) system to the processing in the computer-aided manufacturing (CAM) system to create what is now called a "CADCAM" integrated scenario. In this CADCAM integrated scenario, the operator can call up the specific part program geometry from a specific geometry file and then translate that geometry through a series of processing routines directly into cutter location files and ultimately generate the tool path automatically in a graphics portrayal.

Several automatic machining routines have been created that allow us to translate these frequently used part programming routines into specific numerical control languages automatically. These are called *macros,* where the macro subroutines have been developed as a part of the CADCAM system software. Almost all CADCAM system vendors now have macro routines that allow one to translate from geometry for a part product into geometry for process in a consistent and rigorous fashion.

The second recent advancement in numerical control is the opportunity to eliminate the translation between various process development steps. Typically, using conventional numerical control, we have encountered many problems, such as mistakes in the creation of the part program, incorrect selection of how the process should be performed, inability to maintain quality communication and media support, inappropriate reading and writing onto these tapes, and a myriad of other problems caused by the intermediate Mylar-based methodology. The advancements in technology have eliminated the Mylar tape and allowed communication through an electronic network from the numerical control supervisory controller to a numerical control machine controller with direct linkages.

Two strategies often are espoused to do this. One, using a supervisory controller that links to a single machine tool controller, is typically called CNC, or computer numerical control. However, some companies want to centralize the overall processing of the numerical control developments and make that available throughout the world from a single source. This is often called direct numerical control, or DNC. In this manner, many machine tools can be controlled from a single, supervisory unit.

CNC tends to be one small computer for each machine tool. DNC tends to have one large computer operating many machine tools. That's the differentiation used in this text to represent the two notions. They both utilize the same functionality and perform the same functions. It's simply a question of the architecture used to create that process.

In both the CNC and DNC structures, the ability of the microcomputer or minicomputer to drive the servos and to interface with the logic of the machine tool controller allows the computer to drive the overall operation of the machine and also sense in a closed-loop methodology what the motion is for that particular process.

## SECTION 6.11  CLOSED-LOOP CONTROL

As we've seen in this chapter, true advancements have occurred in the ability to produce a precision part in an integrated fashion and to take geometry profiles for a part as specified in a computer-aided design system and integrate those with machine controllers and methodologies to define how the machine should operate. The only requirement now is to close the loop to ensure that what the machine is actually performing is what has been specified. These closed-loop controllers, sometimes called *adaptive controllers,* are used to sense and then to make minor corrections as the process is being performed. Adaptive control was also developed under support from the U.S. Air Force with contracts to various aerospace companies in the early 1970s. They sought to identify and develop sensors by which machine problems could be detected early and also to close the loop to ensure proper products and processes were being performed.

The unique aspect of the numerical control concept is that it was designed to effectively accommodate one part at a time. No changes are needed in the control of the machine tool to support the 21st century paradigms. However, the integration of these systems is forcing a fundamental paradigm shift to the adoption of worldwide standards to embrace the notion of "globalization with localization." In this way, a company can operate many modules of manufacturing in an integrated manner. This allows efficiencies—

thus, the leveraging of a *global* environment, while still being responsive to the *local* customer with rapid changes to the systems and rapid solutions to customer problems. New tools and standards, such as PDES and STEP, will do much to allow this globalization with localization notion to be achieved.

## SECTION 6.12   CONTROLLING THE INVENTORY AND MATERIAL FLOW

A fundamental to integrated process design and development is the way the inventory and flow of material is handled in the enterprise. All of the sophisticated planning is of no value if the plan cannot be controlled and executed properly. For example, if the plan is to be world class by achieving flexibility and managing to time, yet the controls on inventory and material flow allow large inventories to build through a supply-push system, very little real savings and very little improvement to customer satisfaction will be achieved. Therefore, it is imperative that the inventory be controlled in a manner consistent with our world-class requirements. To achieve efficiency, effectiveness, and flexibility, the notion of demand pull becomes very important. Demand pull is driven from a customer viewpoint with a "look-ahead" philosophy that attempts to meet the customer needs in a minimal amount of time.

Clearly, the scheduling of materials is fundamental to the manufacturing of the product. This is covered extensively in asset management texts. But through the integrated product-process environment defined herein, new elements of the enterprise must also be scheduled and utilized in new, integrated fashions. For example, the scheduling and control of the inventory of tooling now takes on much greater leverage as the notion of group technology and SMED is employed. By leveraging on the similarities and by minimizing the downtime between lot sizes, management of the tooling becomes critical. The second element of the importance of material flow and material control integration is envoked when new materials such as nonmetallics add the dimension of time out of the refrigerator. Finally, the third element of complexity is the effect of the environmental aspects of material control. Companies are now adding hazardous materials management systems to their portfolios to manage and track all materials that have some element of environmental hazard associated with them.

## SECTION 6.13 THE MIGRATION TO INTEGRATED PROCESSES

Now that the basics of process control are clear, a review of how these techniques have evolved from simple nc of the 1950s to the complex systems of today will prove useful. We should also recognize how various cultures look upon this notion of flexible manufacturing systems (FMS). Jaikumar has developed an excellent comparison of U.S. and Japanese applications of FMS that is shown later in this section.[5]

To begin, the timeline shown in Figure 6–11 shows how machining and material flow systems have evolved over the past thirty years. As discussed earlier in this chapter, the concept of using a computer to control the process began in the 1950s and matured in the 1960s. Computer Numerical Control (CNC) and Direct Numeri-

**FIGURE 6–11**
**A Timeline for Integrated Processes**

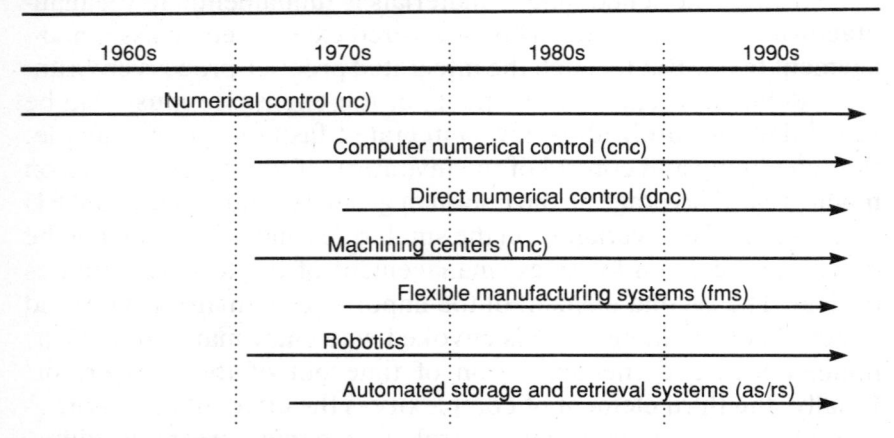

| 1960s | 1970s | 1980s | 1990s |
|---|---|---|---|
| Numerical control (nc) → | | | |
| | Computer numerical control (cnc) → | | |
| | | Direct numerical control (dnc) → | |
| | Machining centers (mc) → | | |
| | | Flexible manufacturing systems (fms) → | |
| | Robotics → | | |
| | | Automated storage and retrieval systems (as/rs) → | |

Note: All dates are approximate, used only to show relative timelines.

cal Control (DNC) were advanced notions of nc that placed a computer in the loop of the machine controller to perform the complex calculations and to achieve the desired repetition.

In the early 1970s, sophisticated machine tool companies such as Cincinnati Milacron, Kearney and Trecker, Giddings & Lewis and many Japanese tool makers began to address the set-up time delays by creating machining centers (mc). These machine centers integrated sophisticated, precise nc "spindles" with tool changers for rapid use of multiple tools, multiple beds for off-line set-ups to insure good utilization, and even multiple spindles to increase the available value-added times on the machine.

Finally in the late 1970s, multiple machines, complete with integrated pallet handling systems, integrated feedback systems, and sophisticated scheduling systems, were introduced. These were called flexible manufacturing systems (FMS). But the term *flexible* is one of conjecture. In his research, Jai Jaikumar has shown how the United States and Japan really plan and operate FMS in two different ways. Table 6-1 compares the United States applications of "flexibility" with the Japanese applications. Clearly, the Japanese have developed FMS's more rapidly, yet still achieve greater flexibility. This paradox has led to "remarkable increases in productivity" according to Jaikumar.

**TABLE 6-1**
**U.S. versus Japanese FMS Comparison**

| Manufacturing System Variables | United States | Japan |
|---|---|---|
| System development time (years) | 2.5 to 3 | 1.25 to 1.75 |
| Number of machines per system | 7 | 6 |
| Types of parts produced per system | 10 | 93 |
| Annual volume per part | 1,727 | 258 |
| Number of parts produced per day | 88 | 120 |
| Number of new parts introduced per year | 1 | 22 |
| Number of systems with untended operations | 0 | 18 |
| Utilization rate (two shifts)* | 52% | 84% |
| Average metal-cutting time per day | 8.3 | 20.2 |

* Ratio of actual cutting time to time available.

Parallel to the advancements in machine processes have been advancements in material movement and storage. In the early 1970s the concept of robotics began to take form. A *robot* is defined many ways. The Computer Aided Manufacturing–International (CAM–I)[6] defines a robot as "an automatic apparatus or device that performs functions ordinarily ascribed to human beings or operates with what appears to be almost human intelligence."

The Japanese tend to use a looser definition that qualifies any automated material movement device—thus the many disparities in the 1980s when comparing robotic installations. Whatever definition is used, the notion of automated material handling is critical for a variety of tasks:

- Unsafe or tedious environments.
- Criticality of movement precision.
- Speed of movement.

A "copy-cat" situation in the early 1980s created a dilemma when one U.S. company simply bought a robot to match another company's purchase. The applications were often to mimic a human and many failed, thus affecting the robotic concept. Since that time the notion of simplify, then integrate, and finally automate has taken on new meaning. Companies now automate only the "right" task, and then only for additional savings in cost or time or improvement in quality. It is the quality of the repeatability of the robot that will dictate its future use into the 1990s.

A second integrated material handling concept that arose in the 1970s was that of an "automated storage and retrieval system" (AS/RS). Clearly, if a lot of inventory is to be held, then a rigorous methodology for storing and retrieving this inventory is warranted. But a second material handling dilemma arose here as well. To run manufacturing in what is now called a "lean" state, meaning demand-pull of product with short-cycle manufacturing and minimal inventory, an AS/RS installation must be questioned. Clearly it adds rigor to inventory management, but why keep the inventory at all? What is evolving is a total rethink of how rigor can be achieved in this "lean" state. The notions of mini-AS/RS's now can be justified to manage the flow into and out of FMS. But the mini-AS/RS must be the storeroom for most components used, including the work-in-progress (WIP), the perishable and non-perishable tools and fixtures, and the inventory of spare parts to insure proper equipment reliability. With the AS/RS becoming an integral part of the streamlined cellular manufacturing design, rigor can be achieved.

## REFERENCES AND RECOMMENDED READINGS

1. Richard C. Dorf, *Modern Control Systems* (Reading, Mass.: Addison-Wesley Publishing Co., 1967).
2. David D. Bedworth and James E. Bailey, *Integrated Production Control Systems* (New York: John Wiley & Sons, 1987).
3. Mikell P. Groover and Emory W. Zimmers, Jr., *CAD/CAM* (Englewood Cliffs, N.J.: Prentice Hall, 1984).
4. Harry D. Moore and Donald R. Kibbey, *Manufacturing Materials and Processes* (Homewood, Ill.: Richard D. Irwin, Inc., 1965).
5. R. Jaikumar, "Post-Industrial Manufacturing" *Harvard Business Review*, vol. 64, 1986.
6. Computer Aided Manufacturing–International, Inc., *Glossary of Computer Aided Manufacturing Terms* (Arlington, Tex.: CAM–I Press, 1978).

# CHAPTER 7
# PROCESS METRICS

We have focused on integrated process design and development from a variety of viewpoints dealing with the prerequisites for the process, the planning of the process, and the control of the process. Now we introduce how one monitors and measures these processes and how one designs an integrated process design and development measurement system that can meet requirements in the 21st century. This will be called the "process metrics."

## SECTION 7.1 TRADITIONAL METRICS

Traditional metrics are often described in a tongue-in-check fashion as "new, good, fast and cheap," where *new* responds to the ability to create new products on a timely basis, *good* represents the ability to meet customer requirements from a quality standpoint, *fast* represents the ability to get products to market in a timely fashion, and *cheap* means the return on that asset and the cost of goods sold provide a margin by which the price minus the cost can lead to profitability.

In traditional metrics, new is often measured with respect to the ability to translate the research and development expenditure into the creation of new products in a timely manner. Good has been measured in terms of upper and lower control limits and acceptable quality levels as defined by X-bar, R charts and traditional measures of quality control. Fast has been measured by the ability to respond in a timely manner and capture market share before the competition, and cheap has normally been developed as a return on net asset calculation using the traditional metrics developed in the 1920s in the du Pont model.

But the universal metrics are changing. In the first chapter of this text, Drucker states, "We must provide significantly new metrics to achieve significantly better visibility." For example, we must align the information systems with the accountability; we must provide rigorous and reliable feedback; we must break the traditional direct labor measurement standards and identify not only the *cost of production* but also the *cost of nonproduction*. And we must embed the physical process into the economic process. If we look at these as the baseline by which we are going to develop the metrics for the 21st century manufacturing community, the new metrics are evolving and that's what we will discuss next.

## SECTION 7.2  CHANGING STRATEGIC DIRECTIONS

In 1980 and again in 1985, Michael Porter from the Harvard Business School developed a set of theories that addressed how products could compete.[1] He introduced the notion of generic business strategies. These are based on his idea that some structural uniqueness is a necessary condition for an enterprise to have a long-term, sustainable competitive advantage. During the early 1980s, market share was deemed the predominant metric by which product success could be measured. But Porter warned, "This approach is as dangerous as it is deceptively clear. While market share is certainly relevant to competitive position, industry leadership is not a cause but an effect of competitive advantage." He hypothesized that an enterprise can possess two fundamental types of competitive advantage:

- Low cost.
- Differentiation.

And in each, an enterprise can seek two different scopes for the target market:

- Industrywide.
- Specific segment.

This is shown graphically in Figure 7-1.

But today, new notions being used for integrated product-process design and development are supplementing the generic strategies of Porter. These can be characterized as:

- Core competencies as developed by Prahalad and Hamel.[2]
- Time-based competition as developed by Stalk.[3]
- Strategic alliances.
- Continuous improvement, as developed via TQM.

**FIGURE 7–1**
**Porter's Model for Generic Strategies**

|  | Strategic advantage | |
|---|---|---|
|  | Uniqueness perceived by customer | Low cost position |
| Industry-wide product offering | Differentiation | Overall cost leadership |
| Particular segment only | Focus | |

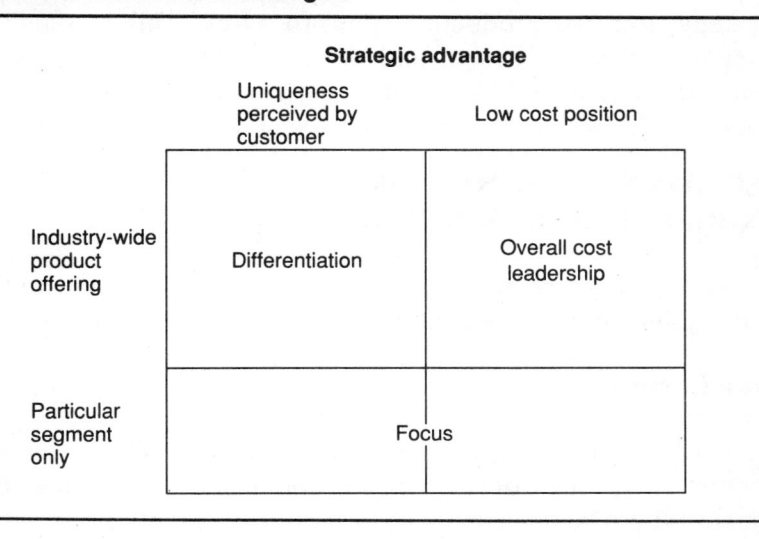

**FIGURE 7–2**
**Competitive Strategies Matrix**

| Porter's classic strategies | New product- and process-oriented competitive strategies | | | |
|---|---|---|---|---|
| | Core competency | Time-based competition | Strategic alliances | Continuous improvement |
| Differentiation | | | | |
| Low cost | | | | |
| Focus or global | | | | |

Each of these supplemental strategic drivers plays a role in the creation of strategic systems and products. In this text, the focus is on the new, supplemental strategies because the focus of the text is on "process." However, we cannot overlook the impact of "product" on the competitive positioning of the enterprise. Figure 7–2 attempts to capture in matrix form both Porter's ideas on generic strategies and the new, enterprise strategies listed above. In this way, a segue can be established in the creation of a corporate strategy from the products to the processes. This will lead to an integrated process design and development technology strategy that is directly linked to the business strategy.

## SECTION 7.3  THE NEW METRICS FOR WORLD-CLASS COMPETITION

As we look at these new strategic ways to compete, we are finding that significant changes exist in the marketplace of today.

### Core Competencies

The strategic directions of the world's major manufacturing economies are migrating from a financial capital metric to a human capital metric, where human capital is the ability of the enterprise to exploit and to tap into the wealth of knowledge that exists within the people of its organization. This implies that physical work is migrating to knowledge work and that few people will be touching the product in the 21st century, although many people will be adding value to the product from a knowledge standpoint.

### Time-Based Competition

For this to succeed, time must be accurately measured, and it must serve as a strategic driver to many corporate decisions. Boeing has invested huge amounts of capital dollars in its ability to maintain the configuration of every airplane it has produced in order to provide excellent service to its customer whenever an airplane needs a spare part. Although the cost of this configuration system may not have a direct payback using Porter's model, it has a tremendous payback when directly linked to service to the customer.

### Strategic Alliances

By taking the enterprise view, the breadth of knowledge and capability needed by a single business to have excellence in all areas of vertical integration of the processes, total integration of the systems, direct linkages to distribution channels, and consistent understanding of the real customer needs has proved almost impossible. What is evolving is the notion of strategic alliances that foster "enterprise teamwork" as compared to previous strategies that attempted to foster multiple sources for external support so that "competitiveness would foster cost reductions." The shift from competitive posturing to teamwork alliances is again changing the way integrated process design and development is viewed. To be able to link with an aligned partner anywhere in the world is forcing "standardization of systems."

**Continuous Improvement**

In the past 10 years, the significance of making small, incremental improvements has become integral to the success of the corporation. People and technology must be in sync for this to happen, and each must be treated as a valued asset. All of the individuals in the corporation are responsible for continuous improvement, and the corporate culture must be shaped to allow this to happen. In the past, we made technology leaps and paid little attention to the refinement of the process that achieved the success and the redefining of the tasks needed for this success. An example of this is the success rate for Class A installations of MRP systems. Now the focus is on ongoing, smaller improvements versus the big bang. In *Fortune* magazine, Dr. Toyoda of Toyota Motors was quoted as saying his company's corporate objective is to "inch rapidly" toward success.[4]

For continuous improvement, the enterprise must address rigorous process improvement. The migration is from test and test and test to determine if the product meets the customer's requirements to characterization (or structured testing) that is being developed to meet the stringent quality needs of the customers. Test has the connotation that we will make a product and then test it to see if it's correct, whereas characterization looks at the integration of the testing activities early in the process development cycle to fully characterize the product and the process to ensure that every product being produced is a good product. This has a dramatic effect when we recognize that the low technology emphasis of the previous years is migrating to a much higher level of technology, not only in terms of the process but also in terms of the information technology used to support the process.

This has caused a series of paradigm shifts. First, many companies are recognizing there is a service orientation for everything, and the job of everyone is to make his or her customer successful, where the customer is the next one in line. Second, time to market is now not just a metric, but it is also a major corporate asset. If time, based on the Drucker notion, is the fundamental building block for 21st century competitiveness, then time-to-market compression is a fundamental corporate asset that must be exploited. Third, there is a shift to the recognition that the No. 1 internal asset for any organization is the people and the knowledge they possess. Human networking is forcing a major shift in the way corporate cultures are viewing this entire concurrent engineering, integrated product-process development cycle. And finally, there are paradigm shifts introduced earlier that state "slow down to speed up." "Slow down to speed up" is intellectually inconsistent and yet is fundamental to the overall success of the enterprise in a continuous improvement manner if we're comparing the way Western companies manufacture product versus the way Asian Rim companies do.

There are some rules of thumb that deal with these paradigm shifts. For example, in the electronics industry, the rule of thumb is that the first two companies to get a next generation product to the marketplace often capture 80 percent or more of the market. This is a startling realization when it comes to where one invests time and dollars. Also, as companies migrate through the life cycle of a product, fully 70 to 80 percent of that product's cost has been dictated and been determined by the time we leave conceptual design of the product. These two startling facts have led to some shifts in the metrics for the 21st century.

The new metrics not only look internally to the organization, such as research and development expenditures, acceptable quality levels, market share, return on net assets, and cost of goods sold, but they also begin to add a new dimension to the "new, good, fast, and cheap" metrics of today. Responsiveness is the new measure for *new*. It can be measured in many ways. But in the time domain, responsiveness is measured as the *mean time it takes to respond to a changing customer demand*. This "MTTRCCD" metric is an external view that takes into account when a customer expects something and how long it takes the enterprise to respond to that changing customer demand. As the mean time to respond to changing customer demand approaches zero, we near world-class status.

We can also address responsiveness in terms of not just reviewing the utilization of existing metrics but also adding new dimensions to our metrics. For example, the inventory of assets is often used to measure the return, as we discussed earlier, in terms of asset turns. However, the migration to an *inventory of knowledge* is a fundamental new look in terms of how one turns over the knowledge within an organization and how rapidly new ideas are incorporated and accepted by the organization. The inventory of knowledge turnover is just as important in the 21st century as the inventory of asset turnover.

The second idea is that *good* now is not simply measured by acceptable quality levels. To truly address world-class quality in the 21st century metrics, we are dealing with notions such as six sigma, a world-class standard by which we have no more than 3.4 errors per million of opportunities for errors. This allows us to achieve a Cp of two and allows us a very high probability that the product presented to the customer is of world-class status.

From a *fast* domain we are realizing that the traditional metric of market share capture is a reactive metric and certain proactive metrics, such as *time-to-market compression and the ability to respond rapidly to changing customer demands,* are much better predictors in terms of the overall ability to respond to a customer demand.

Finally, as we talk about *cheap,* we're talking about assets and asset utilization and the inventory turns related to these asset utilizations. Traditional metrics have shown that an inventory turn of 10 or 20 may be acceptable when one calculates the inventory turns as simply:

Inventory turns = Total sales / Average inventory

which means that in the 250 or so days that most companies work, there are typically 25 days' worth of inventory in stock at any one time. In today's world, if we have inventory turns of 50 to 100, we are approaching world-class status in many organizations.

In each of these areas, the cost of nonproduction is being highlighted as much as the cost of production. Lack of access to knowledgeable people, lack of the availability of talent, lack of response to changing customer demands, only acceptable quality management, poor time to market, and improper asset utilization are predictors that the metrics must change to reflect modern strategies. New metrics will pinpoint the causes of enterprise problems. These new metrics will show us where we are spending time and money and where we are losing quality as true costs of nonproduction. Our objective is to provide a much greater visibility for management decision making than our traditional metrics, which typically deal only with the costs of production.

## SECTION 7.4  MEASURING THE EFFICIENCY AND EFFECTIVENESS OF MANUFACTURING

To address the efficiency and effectiveness of the manufacturing environment, we must understand a simple equation:

MTT (manufacturing throughput time) = N (the efficiency) × the summation of each of the individual processing times, where processing times are for 1 through $m$ individual processes on the critical path.

$$\text{MTT} = N \times \left( \sum_{i=1}^{m} \text{PT}_i \right)$$

Today's focus has been on the individual processing times on the individual pieces of equipment. Manufacturing, through significant improvement focusing at the process level, has been able to reduce processing times to about the limit. However, given MTT = N × the summation of each of these processing times, N is the dominant factor in this equation. N is the efficiency by which we transfer these processing times into the overall cycle time. N reflects how efficient our setups and teardowns are; N reflects the overall production planning and control efficiency; and N reflects how integrated our material flow system is.

Tomorrow's focus will be on total manufacturing cycle time. In many cases today, we find that the efficiency we have for N is often

100 to 1,000 times the factor of the summation of the individual machine processing times. Hence, it may take us only two or three hours to fully process a product, but it may take two weeks to two months for the total manufacturing cycle time to get the entire process complete. This efficiency notion of N must be attacked if we're going to be successful in achieving world-class standards in the 21st century. An N of 1.1 may be acceptable by today's standards for automobile assembly. That is the factor presented from the new Honda NSX plant in Japan where the summation of the processing times is only 10 percent less than the overall manufacturing cycle time it takes to produce a new automobile.[5] Certainly this is world class by any measure.

This is one of the measures of the manufacturing facility. It focuses on the efficiency of value being added to the product. By establishing a time-based value-added measure, the facility alternatives can be compared. This will be called the "value-added velocity." By optimizing the value-added velocity, the enterprise can achieve significant benefits. Clearly, this has direct relationships to traditional metrics of cost per square foot but, more importantly, it allows a direct relationship to the value the facility adds to the customer.

One way of assessing the value-added velocity is to employ the notion called "A $\Delta$ T."™ This is an intellectually simple technique espoused by Digital Equipment Corporation and others to address the efficiency by which value is added. Step one is to determine the "as is" condition of value added—the A. This is done by creating a process flow model for the products in the facility. This can take the form of a series of linked boxes, each of which represents a function performed in the manufacturing life of the product. Then measure the time or the value that each function gives to the product. By adding all of the times, we can create an "as is" time model; by adding all of the values, we can create an "as is" value model. The A is the actual, the "as is." Step two is to determine the T for the "theoretical best" way of adding value. This is done by assessing the value-added efficiency at each of the functions in the process flow diagram. What the difference is between the A and the T is the delta ($\Delta$). By using this simple technique, a plan of action can be created that will achieve the optimum value-added velocity.

## SECTION 7.5  THE LEARNING CURVE NOTION

The learning curve is a well-known concept relating the performance of a specific task to the number of repetitions of that task.[6] Using some indicator, such as direct labor hours per unit of output, a curve can be drawn similar to that shown in Figure 7–3a and linearized using a logarithmic scale as shown in Figure 7–3b.

Calculations can be made as to the rate of the slope of the curve, thus characterizing it as a "constant percentage of reduction" by using the formula:

$$\text{Time for future item} = (\text{Starting time}) \times (\text{Cycle time})^{-\text{Slope}}$$

For example, the slope of the linearized curve shown in Figure 7–3b is 0.34. With this as the negative exponent and the starting time of 260 time units for the first item, the time allowed for the 60th item would be:

$$\text{Time for 60th item} = 260 \times 60^{-.34}$$

or the value would be 64.6 minutes using the classic formula.

But the enterprise is changing its paradigms, changing the technologies for processes, and creating new manufacturing capabilities. These are dramatically affecting the manufacturing strategies and allowing the enterprise new ways to compete in world markets. New strategies, new organizations, and new management styles are leading organizations to be able to produce the "engineered product" for the changing customer demands. To do this, the historical usage of learning curves is being downplayed and in their place are three new metrics.

**1.** We are shifting to the ability to *pay for skills* as compared to *pay for longevity* on the job. With pay for skills, either a person is qualified to perform a task or a person is not qualified to perform a task. The learning curve for the individual is minuscule.

**2.** By using the six sigma notion for product and process characterization, the use of designed experiments (a structured methodology of testing that allows the minimum tests to be run and still attain significant information on the product or the process) conducted in the past is allowing few, if any, wasted parts to be produced early in the life cycle of the product. The enterprise is recognizing that lot sizes are becoming much smaller and that the

**FIGURE 7–3**
**A Learning Curve Example**

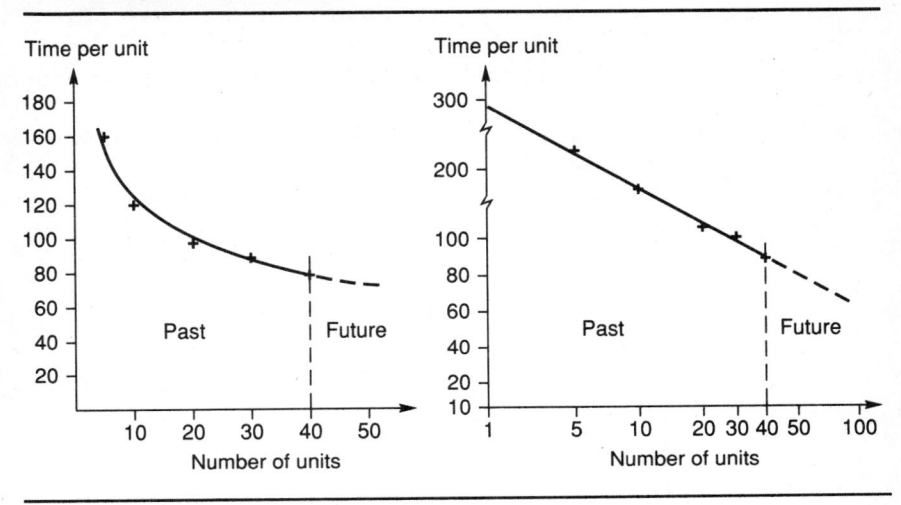

first parts off the line must be good parts for the business to succeed.

3. The utilization of integrated information systems, such as computer numerical control (CNC), to support world-class manufacturing is allowing us to download on a timely basis process plans and process characterizations to the extent that we can produce an economic order quantity of one with a setup time approaching zero. The EOQ of one with a setup time of zero is forcing us to again eliminate the learning curve notion and state that the first part as well as the last part will be produced in the same amount of time. What this allows us to do is get consistency in the metric of time as we migrate to the 21st century manufacturing environment, allowing us to achieve an efficiency factor in our equation MTT = N × the summation of processing times. N now becomes very, very efficient—approaching one.

## REFERENCES AND RECOMMENDED READINGS

1. Michael Porter, *Competitive Strategy—Techniques for Analyzing Industries and Competitors* (New York: Free Press, MacMillan, 1980).

2. C. K. Prahalad and G. Hamel, "The Core Competence of the Corporation," *Harvard Business Review,* May–June 1990.
3. George Stalk, Jr., "Time—The Next Source of Competitive Advantage," *Harvard Business Review,* July–August 1988.
4. Alex Taylor III, "Why Toyota Keeps Getting Better," *Fortune,* November 19, 1990.
5. Roger Schreffler, "Honda's Experimental Assembly Plant," *Automotive Industries,* February 1991.
6. Shlomo Globerson, *Performance Criteria and Incentive Systems* (Amsterdam: Elsevier, 1985).

# CHAPTER 8
# GETTING STARTED IN INTEGRATED PROCESS DESIGN AND DEVELOPMENT (A NATIONAL SURVEY AND TWO CASE STUDIES)

National Science Foundation studies and corporate studies have tried to ascertain how integrated product and process design and development can be achieved and their implications to the overall factory. Efforts to streamline the planning, control, and execution of integrated product-process design are being dealt with in three dimensions (as found in a National Science Foundation study under Grant ISI85-00496).[1] The most important results were:

1. Administrative experiments for streamlining used to support integration of the organization's hierarchy yielded an average throughput time reduction of 54 percent for new manufacturing systems.
2. New administrative practices used to integrate with component and material suppliers are significantly correlated with the percentage of target cycle time achieved and inversely and significantly correlated was scrap and rework.
3. Administrative experiments to promote closer coordination and response to customers during modernization were significantly and inversely correlated with changeover time, with an average of six hours.

This survey, by Dr. John Ettlie of the University of Michigan and others, concludes the weakest link yet to be sufficiently appreciated by business units is marketing-manufacturing. This coordination brings the end customer onto the factory floor with quality, delivery, and service concerns. We know less about this link and, more generally, the source of administrative innovations in domestic firms than anything else, Ettlie concludes.

To address this, another NSF research study has been launched linking the University of Michigan, Vanderbilt University, and Arizona State University to identify and define the overall management of the product-process design systems in manufacturing.[2] The study's intent is to form a team of engineering, business school, and industrial researchers and practitioners to demonstrate a concept of integrated, *optimized* design in a U.S.-based cultural setting. Because of the nature of the project and its significance, it is being conducted in two phases: Phase one is the demonstration program, and phase two will be the proprietary implementation phase. The results of phase one will be displayed here as the beginning notions of Chapter 8 as we begin to identify what is necessary to achieve integrated process design and development in manufacturing.

The reasons for this study are quite simple. As two recently concluded National Research Council studies have shown, the basis for a technical framework is needed to achieve design optimization.

The first of these two studies is the National Research Council's joint effort with the National Science Foundation titled, "The Competitive Edge Research Priorities for U.S. Manufacturing." During the formulative stages of this study, the team identified that an integrated product-process realization framework was mandatory for success. Figure 8–1 shows a "Computer-Aided Product Realization Framework." It calls for a technical system to create a consistent, rigorous, transportable image of the product that can transcend all aspects of product definition and product realization. The premise for this is based on the product definition exchange standard, or specification (PDES), introduced earlier in this text.

It is also based on findings of a second National Research Council study on rapid prototyping that was conducted under the support of the Manufacturing Studies Board of the National Research Council. The objectives of the studies are to conclude and to ascertain which of the typical barriers shown in Table 8–1 are present and how one can overcome these barriers to concurrent engineering implementation throughout the industry.

In Section 8.1, we will begin by summarizing the results of the preliminary National Science Foundation's survey of American manufacturing companies as they begin to address integrated prod-

**FIGURE 8–1**
**Computer-Aided Product Realization Framework**

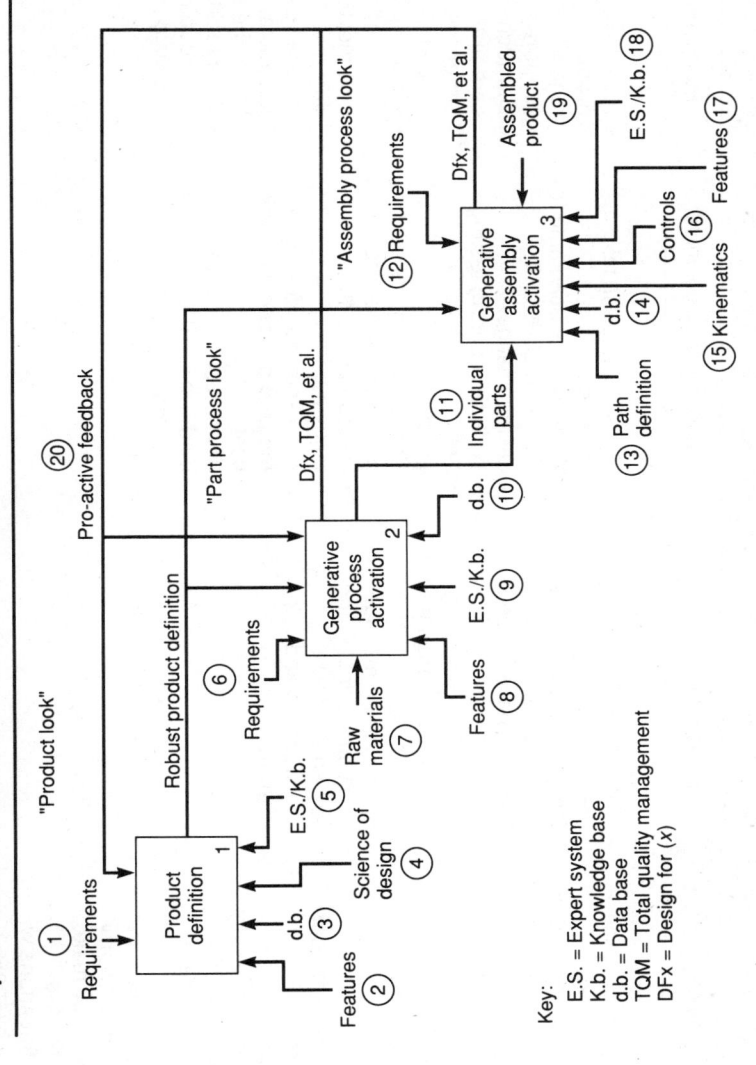

## TABLE 8–1
### Typical Barriers to Concurrent Engineering

*Program Phases*

| Concept Formulation | Validation | Full-Scale Development | Production Deployment |
|---|---|---|---|

*Managerial*

| | | | |
|---|---|---|---|
| • This means changing the way we do business—we're already struggling to stay even with the competition | • Let's give more budget to design | • Our suppliers will take up the slack | • Have to meet our production quota regardless of supportability |
| • We've done without CE and have always been successful | • Our baseline design is fundamentally good | • We've always been able to fix it with ECPs and design changes | • Our technicians will take care of the field problems |
| • Too soon to allocate budget for CE we don't have anyway | • Let's not have CE be a show stopper | • We'll fix it during testing | • Other systems are having problems as well as ours |
| • Let's give it some support for the record | • Tell them to wait until detail design, then we'll consider it | • Let's push for early production | • We don't need a feedback loop for our future designs, they'll all be so different |
| • Our experienced designers can handle any challenge without CE | • Our program manager must cut budget—and you know where | • Can't allow any impact on production schedule | • It's the customer's technicians that always break the system |
| • Let's get a committee started | • We stand behind our designers | • We don't need these fancy IMSs | • We'll tackle each problem individually as we've always done |
| • Our production lines are too small for us to worry about CE; we only make oneses and twoses | | • Let's keep our designers focused on schedule and cost | |
| • Detailed justifications will be necessary by those "technocrats" | | • We're just too busy with all these CDRLs and other reports | |
| • Does the customer really know what he/she wants? | | | |

*Technical*
- Our designs have always been producible and supportable
- Now they'll tell us how to design it or drop their standards on us
- We'll look over the shoulders of our young designers, that's all that's needed
- Do we really need all these data bases this early?
- We are just doing trade studies—CE is not needed
- The customer doesn't really care about CE—performance counts
- Lessons learned don't apply to this program or its design
- We already have so many specs to worry about
- We've always designed it to be producible, although this material is a little tricky to fabricate
- We will have to cut weight to meet performance—producibility and supportability won't be that adversely affected
- We just don't have all the data
- Those "-ilities" with that jargon can't be understood
- If it's producible it will be supportable
- Those design changes won't affect producibility and supportability
- Those CE cultists are always pessimistic
- We can select the cheapest component to cut costs
- Just no more weight increases
- Those early designs didn't address these issues—too late now
- If it breaks we'll fix it with the ECP or they'll just have to put up with it
- Those predecessor problems weren't caused by the design
- This technology doesn't need CE
- Too late to change, we're in production
- The next design will be done by someone else anyway
- Those "-ilities" have to blame someone for all the production and support problems
- Integrated weapon system data bases are too ambiguous and never provide all the information we need
- It's just a minor problem, otherwise it meets all the specs
- Early feedback data is always bad
- It wasn't intended for that use anyway—no wonder they broke it
- Their skill levels are just too low to fix our delicate widget—they'll come up on the learning curve soon

uct and process design and development. The second major aspect of Chapter 8 will be the integrated approach to competitive designs, looking at a structured approach from a major American manufacturing company that has been willing to provide detailed analysis of where it is heading with respect to world-class competition. The third major element of this chapter is a status report of the implementation of these integrated product and process design principles at a major aerospace company in Phoenix.

## SECTION 8.1 SUMMARY RESULTS OF THE NATIONAL SCIENCE FOUNDATION SURVEY FOR INTEGRATED PRODUCT DESIGN IN MANUFACTURING

This survey conducted early in 1991 surveyed 43 manufacturing companies and divisions to determine the state-of-the-art product and process development philosophies, strategies, and practices in U.S.-based manufacturing. Respondents were usually middle managers or top managers, 63 percent and 25 percent, respectively. Although 31 product types were represented, most were found to be in categories SIC 34 through 37, which are fabricated metal products, machinery, electrical equipment, and transportation equipment. Most of these are from the aerospace and automotive industries. Twenty-four of the responding firms were in the Fortune 250, and the sample of respondents came from divisions of corporations that averaged $16 billion in sales.

Results were unexpected on several issues. Although all but one of the responding firms said they had a program to upgrade the way they developed products, only a slim majority of these companies, 24 of the 43, or 56 percent, actually benchmarked on product development processes. What is more, manufacturing almost never gets *formally* involved in the concept development stage, yet this may be the most important part of the product development effort.

Comparing survey results from 1991 with another study in 1987 is interesting. While companies were using more training and development in 1991 than they were four years earlier in design for manufacturability to enhance their design practices, there still appears to be resistance to adopting job rotation among functions as a major way of doing this, although this result differs from a similar survey of Indiana industry. Marketing is still the No. 1 choice to participate in the concurrent roundtable outside the engineering organization, and quality is second, as was found in the 1987 survey.

Firms continue to increase their use of manufacturing sign-off and new structures (e.g., teamwork and team building) to implement their new product development philosophies. There continues to be a gap, however, between degreed design engineers and degreed manufacturing engineers. The average percentage of degreed design engineers in the sample was 75 percent while the average percent of degreed manufacturing engineers was only 61 percent. The average ratio of design to manufacturing engineers was 4.69 design engineers to each manufacturing engineer, whereas the ratio was 2.88 to one in the 1987 survey. Hence, the equalization of design engineering and manufacturing engineering appears to be going in the opposite direction from the desires and the intents in the statements made by many manufacturing organizations.

In 1987, 13 percent of the sample used in-process measures; whereas in 1991, 35 percent used these same or newer measures. Not surprisingly, respondents believed the most important determinant of success in product development was to *know the customer* and to *integrate the functions*. What is more, the *absence of good communication* and *goal clarity* were the most frequently mentioned barriers to achieving high quality designs in a very short time. Considerable interest in product-process development has been generated by successful firms in the 1980s by their use of benchmarking. Xerox, for example, has been widely celebrated along with many other successful firms for their use of measuring and comparing various dimensions of world-class manufacturing product and process practices to help improve performance and quality.[3] Xerox benchmarked for costs, methods, and service in the 1980s. The NSF study assumed benchmarking would be an integral part of changing the philosophy of product development in other domestic manufacturing firms. When 98 percent of our preliminary survey respondents said they have a program to upgrade product development, the survey also assumed a majority of these same responding firms would also be benchmarking. However, *this was not the case*. In response to the question, "Do you benchmark?" the survey found that only 24 of the 43, or 56 percent, of the respondents said yes. That is, only about half of these firms benchmark in product development, even though nearly all are changing the way they do product development.

As would be expected, based on these results, about half of the respondents named a benchmark firm as the target for product development. Nine of the first-mentioned benchmark firms were Japanese. It may well be that many of the responding firms have just embarked on their new product development programs.

Finally, two more questions on the 1991 survey tended to summarize many of the issues surrounding benchmarking for product development. The first asked what was the purpose or goal of the design approach being used in the firm. The most frequently mentioned first goal was the attainment of multiple objectives, especially the maintenance of or reduction of cost and improvement of quality, while being able to get to market quicker with new products. An example was reduced time to market, improved first-time product quality, and minimized product evolutionary changes. A total of 16 respondents, or 43 percent, emphasized the attainment of multiple, simultaneous goals. Interestingly, cycle-time reduction alone was not the dominant product development objective for this sample.

The second most frequently mentioned first purpose was achieving an integrated design approach. This was reported by 14 respondents, or 38 percent. This included not only getting design and manufacturing involved but other functions as well, and it echoes the responses to structured questions covered earlier.

One of the assumptions of the research project was that firms do not do detailed costing studies of new products until they go into production. However, the study also assumed that as part of the adoption of new philosophies of product development in domestic manufacturing firms, these practices might be changing. That is, perhaps some organizations were beginning to drive cost estimation upstream in the new-product life cycle, perhaps even to the point of doing some cost estimating during the design stage.

The study found that 38 of the 43 respondents, or 88 percent, responded yes to the question of whether or not financial planning during product design includes projections of future costs and revenues. The second startling result was that 41, or 95 percent, responded yes to the question "Do you estimate the production cost impact of alternative product designs?"

Allowing the respondents open-ended questions on the mail survey allowed greater depth in certain areas. Key communication issues and integration issues were highlighted. Focus on the customer and quality received the highest number, followed by integration of the functions. Communication and goal clarity were mentioned. Also, the time issues of cycle time and time-to-market compression and customer knowledge were often noted but found lower in the overall assessment.

The survey found that of the 43 manufacturing companies and divisions that responded to the state-of-the-art product and process development strategies, a goodly mix of middle managers and top managers responded from relatively large divisions, and the results were unexpected on several issues. Although all but one of the responding firms said they had a program to upgrade the way they developed products, only 56 percent of these companies benchmark on product development processes. What is more, manufacturing almost never gets formally involved in the concept development stage, yet this may be the most important part of the product development effort. Cycle-time reduction was not only conspicuous by its absence as a driver in product development strategy, but it also was often mentioned as a barrier. Merely including time as a trade-off seems inconsistent with the attention this variable is receiving in theory and practice.

While companies are using more training and development to enhance their design practices, there still appears to be resistance to adopting job rotation among functions. Marketing is still the No. 1 function outside of engineering for involvement in efforts to integrate design, and quality is second. Firms continue to increase their use of manufacturing signoff and new structures to implement their new-product development philosophies. Not surprisingly, respondents believed the most important determinant of success in product and process development was to know your customer and integrate the functions. What is more, the absence of good communication and goal clarity were the most frequently mentioned barriers to achieving high quality designs in a quick time.

## SECTION 8.2  A CASE STUDY OF A WORLD-CLASS MANUFACTURER AND THE INTEGRATED APPROACH TO COMPETITIVE DESIGNS

Getting started is a very difficult task as one progresses into the integrated product and process design and development area. "If a company is not thinking about how to develop new products twice as good, twice as fast, with half the resources, they probably don't have the right mental attitude to effectively challenge global competition."[4] This is a quotation from a world-class manufacturer who has continued to wrestle with how one gets started to become best in class and to compete on a worldwide basis for world-class success.

The approach this company took was to recognize what the world-class competition really is. This company recognized that three aspects set a world-class competitor apart:

1. A world-class competitor satisfies customer's expectations of perceived value better than anyone else.
2. A world-class competitor generates an adequate profit to rekindle and reinforce the necessary investments to support a world-class facility.
3. A world-class competitor is competitive with anyone in the world on function of the product, quality of the product, price/cost of the product, the services the product and the company can offer to the customer, and the company's responsiveness both on deliveries of the initial product and on the ability to change that product base.

This world-class competitor:

1. Is *close* to the customer and is market-focused,
2. Achieves *continuous improvement*.
3. Is *proactive* with a sense of urgency and fleetness of foot.
4. Is *structured* to establish ownership and commitment to change and making customer satisfaction an element of success.
5. Has an integrated marketing, design, manufacturing, and financial system such that strategies and objectives can be

clearly translated into actions and teams can be created to focus on the product-process integration.

The objective is to focus on the issues necessary to achieve this world-class competition state. The company recognized that the traditional approach to product development is in a serial fashion (see Figure 8–2a). But, in parallel with this, the world-class competitor has also recognized that the traditional manufacturing environment is one of batching, movement to a center, and then waiting between centers in order to produce the product. To become world class requires a rethink to an integrated product-process development team as seen in Fig. 8–2b. Figure 8–3 shows the way the world-class product development cycle must come about. This is what we are sponsoring in this text as we talk about formulating the integrated product development team to address the overall customer requirements for product and process development.

But, just as importantly, to support the integrated product development team, a world-class integrated process development team must focus on cellular manufacturing and the fundamentals we discussed in Chapter 4 that are the prerequisites for success—such elements as U-shaped material flows that allow for cells that achieve flexibility and efficiency at the same time, the just-in-time demand pull, the total quality management (TQM) aspects, and then enterprise integration (EI) tools that are linked so that overall manufacturing competitiveness can be on a world-class basis. This also recognizes that the integration of systems—such as the design systems, the analysis systems, and the manufacturing systems—must be achieved in order to support this world-class manufacturing base.

As we compare Figure 8–2 with Figure 8–3, we find some startling differences. One is the integration of the systems; the second is the impact on the people issues; and the third is the resolution that the customer is critical for the overall success of the organization and that the systems must be aligned to meet changing customer requirements. Also, the integration of design systems and the integration of manufacturing systems can play a very effective part in the overall success of the company. But getting started is a nontrivial task. To get started, one does not simply transform the design organization from serial thinking to thinking in integrated

**FIGURE 8–2**
**Serial versus Parallel Design**

a. Traditional → Design → Prototype build → Prototype build → Prototype build → Engineering release → Process plan → Tooling procurement → Re-process plan → Tooling model →

b. World class parallel

→ Product business plan → Integrated product development team

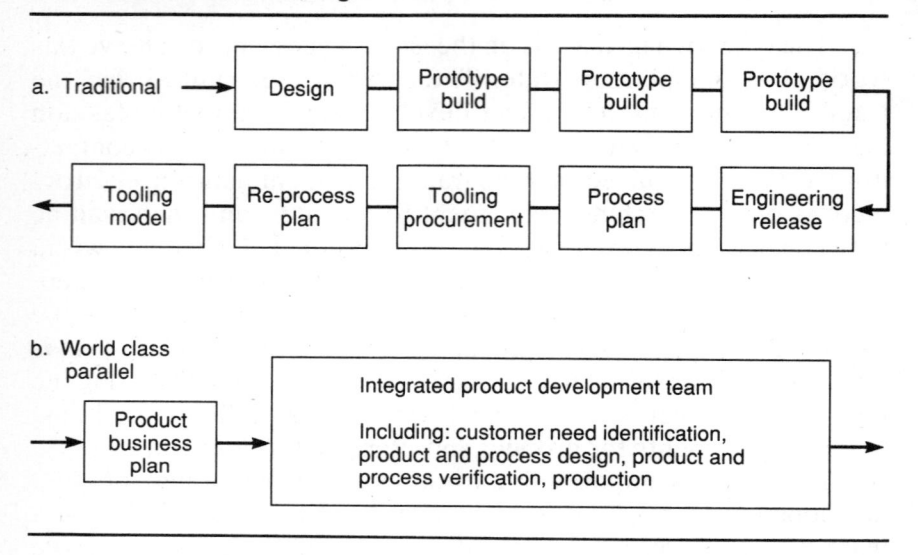

Including: customer need identification, product and process design, product and process verification, production →

**FIGURE 8–3**
**Sequential Manufacture versus World Class**

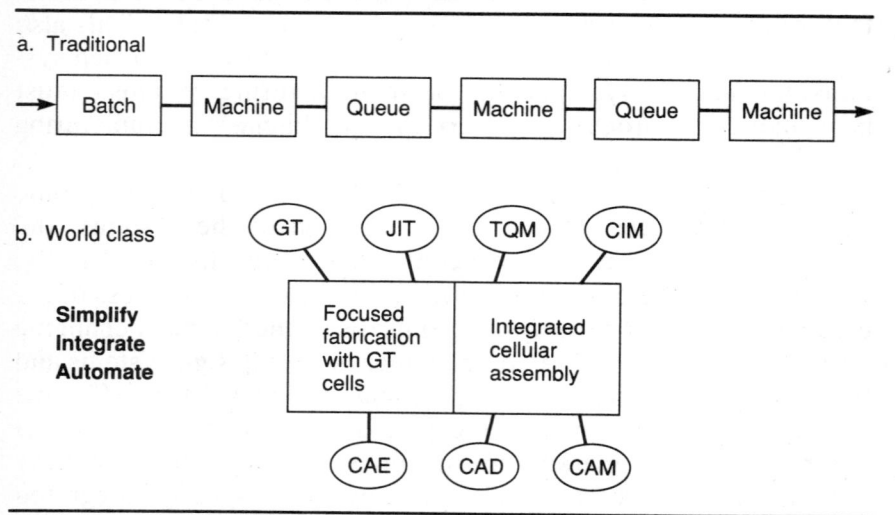

a. Traditional

→ Batch → Machine → Queue → Machine → Queue → Machine →

b. World class

**Simplify**
**Integrate**
**Automate**

GT, JIT, TQM, CIM

Focused fabrication with GT cells — Integrated cellular assembly

CAE, CAD, CAM

product development teams, nor does one rearrange the factory floor to achieve fundamental changes in efficiencies and flexibilities overnight. World-class competitors recommend that companies focus on four elements in design and in manufacturing.

The first element in design is to simplify and streamline the total process. The second is the ability to develop an integrated product development strategy that has the tools in place to translate the customer requirements into the form, fit, and function of the product that ultimately meets the needs of the customer at a quality level that is superior to the competition base and provides a reasonable profit to the company. To do this, the world-class competitor recognizes that the third element is to identify the integration mechanisms that must be in place. And the fourth is to establish a receptive corporate culture for this change.

The building blocks begin with the notion of group technology. The concept is very simple—can we begin to take advantage of similarity and build on the ideas and the knowledge base that we have. It then expands to recognize the customer requirement in a systematic market research environment that assesses not only what the customer is but also where the customer is going, who is the customer's customer, and how can that customer's customer dictate what the customer's success record can be. Very simplistic notions, such as group technology and a true full-fledged customer assessment methodology, can lead to the foundation by which the integrated product development team can springboard.

On the integrated process development side, three critical elements must also be present for world-class status to occur. The first again focuses on simplicity, and the second focuses on group technology. Group technology is the foundation on which the cellular manufacturing basis can be achieved. As we discussed in Chapter 4, group technology simply says, "Can we take advantage of the process similarities by linking together diverse and often nonrelated processes and formulating a production cell" similar to that created at Rockwell and outlined in Chapter 3. By linking diverse processes into an integrated cell, the flexibility and the efficiencies can be achieved simultaneously, and the resultant human factors and teamwork issues account for significantly increasing morale.

The third element is teams of workers that understand not only the product but also the process, and not only individual elements of the process but also the complete integration of the process to achieve a semi-finished or finished product. These elements coupled with new notions, such as demand pull in a just-in-time environment with supplier relationships, allow a simple, straightforward, systematic approach necessary to get started.

In summary, to start the move toward world-class status, four very basic issues must be in place for the factory and the design studios to begin to be linked. First is simplicity and streamlining of the total process. For integrated product design, group technology must be understood and recognized, the systems must begin to be integrated, and a complete understanding of what the *customer* requirements are must be undertaken. From a manufacturing standpoint, again simplicity and group technology, but this time on a process view, must be undertaken to perform cellular manufacturing. Systems again must be integrated, with a demand pull and *supplier* integration recognized as the third element of this success story. These four simple tools for design and three simple tools for manufacturing all are fundamentally necessary for the customer to be satisfied, the supplier to be integrated, and the firm to show efficiencies and flexibilities at the same time. Figure 8–4 demonstrates how this all can be integrated.

One must recognize that training and education are of fundamental importance in achieving this integrated product and process design and development scenario. Training and education must be

**FIGURE 8–4**
**Basics of Integrated Systems**

| Focus | Product | Process |
|---|---|---|
| 1. | Simplify and streamline | Simplify and streamline |
| 2. | Develop integrated product strategy | Develop integrated process strategy |
| 3. | Form teams to identify integration needs (include group technology) | Form teams to identify integration needs (include group technology) |
| 4. | Formally identify customer integrated requirements | Formally identify supplier-integrated requirements |

focused at all levels within the management pyramid, as shown in Figure 8–5. In this figure, the world-class manufacturer has truly focused on the stratification of education and training at four levels within the organization. At the top-management level, overview courses are developed to understand the process, to monitor the process, and to be able to ask the right questions to achieve this integrated scenario. At the program manager level, the process is the focus of the training and education—what is the process, what is necessary to lead the process, and how one can begin to assign the resources necessary to achieve the success.

At the facilitator level, the techniques and the technologies, the tool kit, are introduced to the overall team. The training and education scenario offers the understanding of the process and the ability to coach. Finally, at the team member level, individual work packets and work assignments must be well understood. Education of the whys and the whats migrates to the tool kit of the hows and the how much. What is the process, and how does one perform the work in a very, very efficient and flexible manner to achieve world-class success?

Finally, the benchmarking exercise summarized earlier must be brought into position. In the survey shown earlier, we found that only 50 to 60 percent of the companies in the United States are aggressively pursuing benchmarking exercises. Why? Benchmarking provides insight in terms of where the competition has been, but it gives very little visibility in terms of where the competition is going unless the benchmark is analyzed over a time domain.

One of the objectives is to understand and establish aggressive targets and aggressive metrics that go well beyond the benchmark results. For example, if the current "as-is" condition for the company says a product can be established and go from concept to first-unit delivery in three years, and the benchmark says the "best in class" can achieve that product-to-process delivery in two years, then an aggressive target of one year or 18 months must be established for the company to achieve world-class status. These aggressive, world-class targets must be in place, and everyone within the organization must understand that these are the goals and the objectives of the organization as the organization strives to become a world-class manufacturing organization.

**FIGURE 8–5**
**Training and Education of the Work Force**

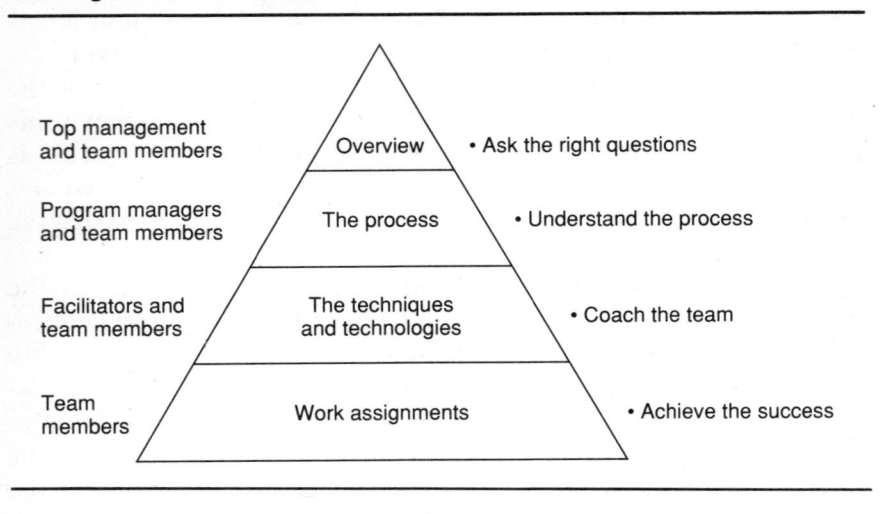

| Audience | Pyramid Level | Purpose |
|---|---|---|
| Top management and team members | Overview | • Ask the right questions |
| Program managers and team members | The process | • Understand the process |
| Facilitators and team members | The techniques and technologies | • Coach the team |
| Team members | Work assignments | • Achieve the success |

## SECTION 8.3  THE STATUS OF CONCURRENT ENGINEERING IMPLEMENTATION AT THE GARRETT ENGINE DIVISION OF ALLIED SIGNAL CORPORATION

In a public presentation, senior managers at Garrett Engine Division of Allied Signal displayed the overall status of how integrated product and process development is being conducted at their company.[5] The presentation began with the recognition that the challenge facing Garrett Engine Division was to implement a *system* that would reduce development costs, meet design to cost the first time, reduce production costs, and streamline marketing, administrative, and support costs to ensure unqualified customer-satisfaction-first attitudes and actions and to reduce manufacturing and procurement lead times. The objective was to move away from the traditional design process, as shown in Figure 8–6, which goes from the design, to the hardware fabrication, to the test, to the production, to the sales order release, and ultimately delivery to the OEM and field use.

This sequential series of events has been traditional at Garrett for many years. The division found that migrating to integrated

## FIGURE 8-6
### Garrett's Traditional Design Process

**Mostly a sequential event**

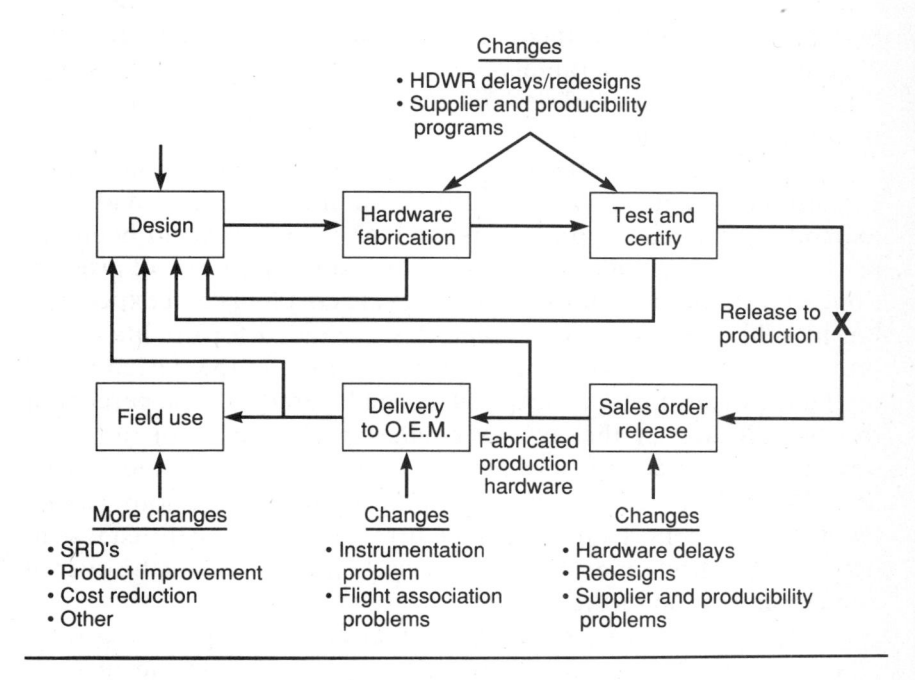

```
                              Changes
                         • HDWR delays/redesigns
                         • Supplier and producibility
                              programs

    ┌────────┐      ┌──────────────┐      ┌──────────┐
    │ Design │─────▶│  Hardware    │─────▶│ Test and │
    │        │      │  fabrication │      │ certify  │
    └────────┘      └──────────────┘      └──────────┘
                                                │
                                                │ Release to  X
                                                │ production
    ┌────────┐      ┌──────────┐      ┌──────────────┐
    │ Field  │◀─────│ Delivery │◀─────│  Sales order │
    │  use   │      │ to O.E.M.│      │   release    │
    └────────┘      └──────────┘      └──────────────┘
                            Fabricated
                            production
                             hardware

  More changes          Changes                 Changes
• SRD's              • Instrumentation       • Hardware delays
• Product improvement    problem              • Redesigns
• Cost reduction     • Flight association    • Supplier and producibility
• Other                  problems                problems
```

product-process engineering required taking a systemic approach to the optimum engine design and manufacturing processes. Their objectives were to integrate the manufacturing processes concurrently into the design of the product and to form teams of manufacturing, purchasing, suppliers, materials, marketing, and product support individuals with the design engineers from the conceptual design onward. The process was intended to cause developers to consider all elements of the engine life cycle from conception to disposal and included such elements as quality, cost, schedule, customer and user requirements and needs.

To do that, Garrett established what it called the "$C^2I$" philosophy for "commitment to continuous improvement," where concurrent engineering was one of many elements the management vision contained. Figure 8–7 encapsulates all the thrusts launched under $C^2I$. The philosophy included such things as the group technology process, the material management process, the tool management process, continuous improvement, quality function deployment, just in time, the ability to understand the cost benefits, the ability to transition rapidly to the production process, and so on. It incorporated all disciplines and concentrated the resources where they were most effective.

A survey at Garrett Engine Division found that, at the end of conceptual design, only 3 percent of the life cycle cost had been expended, whereas approximately 11 or 12 percent had been expended by the time of full-scale development. By the end of production, 30 to 35 percent of the total life cycle cost had been expended, but in field operations 50 percent was expended for the life cycle. However, as one looks at the *leveraging* effect of the design phase on the life cycle costs, conceptual design levered 70 to 80 percent of the overall cost of the turbine engines produced, with full-scale development affecting only about 15 percent, production having less than a 10 percent leveraging, and the field operations having less than 5 percent. Hence, the concurrent engineering thrust was to do it right the first time and affect the early processes and the overall conceptualization of the process.

Three aspects of integrated product-process engineering were addressed by the Garrett Engine Division team: (1) the process definition, (2) the supporting tools needed to achieve this process definition, and (3) the formal procedures and informal practices supported with training and education of all people involved in the overall aspects. Garrett Engine Division used a pilot approach to optimize the integrated product-process engineering process. The division formed integrated product-process engineering committees to define the process and to study what other companies' approaches were. It created integrated product-process engineering steering committees to provide team guidance, and it launched three pilot teams, simultaneously, in June 1990. Garrett also launched then supplier partnerships developed with the teams to link the supplier with the engineering engine design. And finally, training was provided to the teams in the division areas that would be affected by this overall process change.

## FIGURE 8–7
## Concurrent Engineering under Continuous Improvement

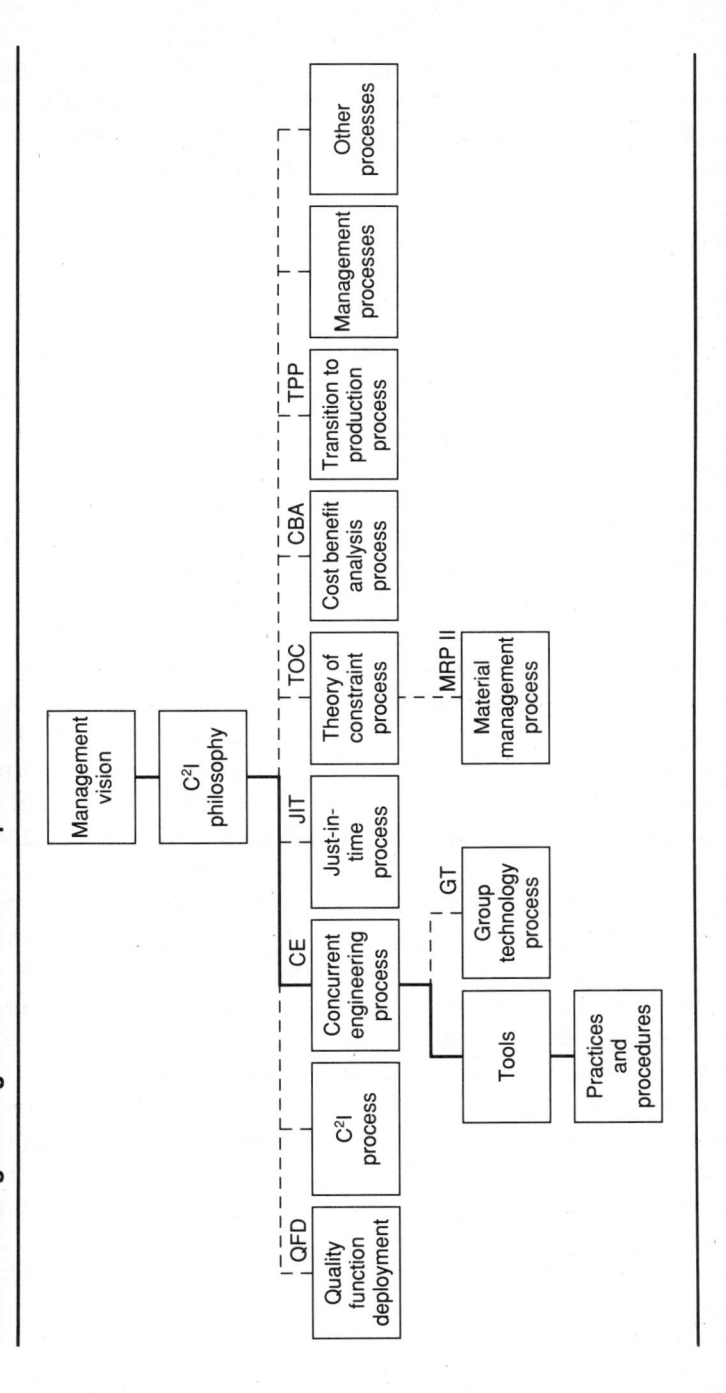

175

One of the targets of opportunity for the Garrett Engine Division was the CFE738 engine. This high bypass modern technology turbo fan engine used on large business jets and 40-passenger commuter applications was a joint venture with General Electric Corporation. The program goals for the CFE738 were (1) a significant reduction of life cycle cost, (2) a weight reduction that allowed the thrust-to-weight ratio to be consistent with modern technology engines, (3) a performance residual that would allow thrust and fuel specifics to be competitive in a world-class environment, and (4) a schedule objective that made it compatible with the customer's needs. Garrett formed module teams, and the program was broken down into logical groupings; that is, engine modules. Each engine module had the team members necessary to support the required functions that would be affected by the engine module. A team leader then organized and directed the module's activities, and assignments would come from within the module, with manpower commitments coordinated with the functional organizations.

The overall process was to move the decision making to the lowest level within the organization. The module teams were found to be too unwieldy to effectively address all the activities, so smaller task teams were formed to address specific requirements. The task teams were made up of intellectual leaders based on the needed disciplines. These task teams selected their own leader, and the team members brought design guidelines from functional organizations to preclude need for management approval. Once these design guidelines had been established, decision making was moved and team findings were presented to the larger module teams, which could then pass judgment. Task teams were chartered for a particular task, and once the task was achieved the team was disbanded.

The integrated product-process engineering approach has been well received at Garrett Engine Division. The teams understand the benefits, but more training and education is needed to be successful on a broad spectrum. Small task teams have opened communication between disciplines; whereas the module teams were found to be too unwieldy to effectively address all these activities. The decision-making process at Garrett Engine Division has been successfully streamlined, and program enthusiasm and commitment has been enhanced.

The second pilot concurrent engineering team dealt with the T800LHT helicopter turboshafted engine. This pilot cost-reduction program dealt with a modern turboshaft engine designed for the Army's light helicopter full-scale development program. This helicopter engine was being scheduled for qualification in 1991 and was a joint venture with Allison Engine Division. The objective of the program was to reduce the cost of the engine.

The team encountered problems: no cost model was available to assist in the decision making, and departments were reluctant to estimate processing costs because the accounting system was not activity based. This led to significant emotional tension involving team members working on another person's design. The lack of familiarity with the engine hardware hindered the input in the creation of this overall cost model. Inadequate attention to organization structure, budgeting process, and equipment before the move was also an initial concern. Also, the team was skeptical that anticipated results could be produced with this cost-reduction effort.

However, the team progressed in focusing on the necessary tasks and created a cost model that brought together individuals from manufacturing, quality assurance, materials, project, and customer support. It included direct, indirect, and all the hidden costs involved in the overall program of the T800. The team built an overall life cycle cost-accounting model that included not only cost values but also time values. The data base status of all the cost suggestions were submitted, and a computer linkage with the time standards department was built such that real standard times could be achieved for each of the processes used to produce the T800 engine. Ideas offering engine fabrication cost savings of 14 percent were identified, and progress was being measured on a significant dollar-reduction basis from the drawings released versus the budget expended.

This second prototyping team learned the following lessons. Program personnel must exhibit excellent people and communication skills. They have to be flexible and able to adapt to new ways of working, and they must be good in their field of expertise. They must have the total support from the uppermost management ranks or the results will not be immediate. Also, the uppermost management teams must realize it will take time to establish a good, sound working team. The initial and ongoing training is substantial and is mandatory. And finally, the people must relocate and co-locate to truly establish synergy. To build team unity, they need to have the sense of working together.

The final of the three integrated product-process prototypes was on the TPF351 turboprop engine. Again, this was an attempt to reduce costs. The TPF351 is a modern technology, pusher turboprop engine designed for the 19-passenger commuter market. It was scheduled for certification in 1991, but it was undergoing this team analysis to significantly reduce the cost of production. This project team has gone through a massive effort to identify the cost-reduction opportunities. Over 242 suggestions have been incorporated to date. The team is also developing what Drucker calls the "hidden cost model," including not only the costs of production of the TPF351 but also the costs of nonproduction. It is also identifying where supplier leveraging can be obtained by establishing concurrent engineering supplier teams.

This third prototyping effort revealed that the move to integrated product-process design and development happens slowly, abundant support is needed from all functional departments, and the teams should report to a neutral person. The upper management team was found to be looking for tangible results much too soon in the concurrent engineering life cycle and did not consider the startup administrative effort required to get this team formulated and on track. The need to teach each member more about how the overall corporation and the division functions as a whole was underestimated.

This prototype also showed the team budget for equipment and travel should be implemented at the inception, rather than have incremental expenditures be justified during the team development. And the manpower and resource budgets should follow shortly thereafter. Other lessons included the need to identify flexible team players and to add team players from other areas. The TPF351 team also learned that a small, core group of 5 to 10 individuals is adequate to start the overall process. However, additional team members may be needed as the project evolves, and access to these individuals is critical for the overall team's success.

Team dynamics necessitate special conditions; a war room needs to be readily available and responsive, and independent secretarial support is necessary. Members need to concern themselves about the reward and the review process, and individual team members must know what will happen to their careers when the effort is over. And they need to understand the impact of reporting to often two different masters, one a functional and the other a team leader.

During these developments, Garrett Engine Division has achieved substantial changes such as six-month lead-time reductions, part reductions from 14 down to 2 on specific compressors, and significant cost reductions. One of the biggest breakthroughs at Garrett Engine Division has been the creation of a plan to make the suppliers an integral part of the team. What has evolved is a dedicated procurement team whose members are enabled and empowered to pursue agendas and be co-located with other disciplined representatives. Suppliers are being brought in during the conceptual stage, and they are being made an integral part of the concurrent engineering team.

In conclusion, the migration to integrated product and process design and development at Garrett Engine Division has been a tremendous learning experience. Good results are coming but not without significant organizational change.

## REFERENCES AND RECOMMENDED READINGS

1. John Ettlie and Henry Stoll, *Managing the Design-Manufacturing Process* (New York: McGraw-Hill, 1990).
2. John Ettlie, "Managing Design Systems in Manufacturing," National Science Foundation Study, Grant DDM-9007043, 1991.
3. Robert C. Camp, *Benchmarking* (Milwaukee, Wis.: Quality Press, American Society for Quality Control, 1989).
4. Sean Battles, "An Integrated Approach to Competitive Design," unpublished, John Deere, 1990.
5. George Davis and Russell Biekert, "Status of Concurrent Engineering Implementation at GED," Chandler, Ariz.: NSF Conference on Concurrent Engineering, 1990.

# CHAPTER 9
# INFORMATION SYSTEM REQUIREMENTS AND DESIGN FOR INTEGRATED PROCESS DESIGN AND DEVELOPMENT

As we develop the notion in this text for the integration of process design and development, we have found certain metrics to be evident if one wishes to conduct process design in the 21st century. New metrics are causing significant new requirements for the manufacturing process. An economic order quantity of one has dramatic ramifications for the way the process is performed and the way the information is needed to support that process. Delays in planning versus execution cannot be tolerated, and volatility in shop floor capabilities and the changes in the shop floor must be reflected on almost a "real time" basis. Because of the expense and the lack of clarity in what "real time" actually represents, this text characterizes the issue of timeliness of data presented to the user as "on-time" basis with respect to those who must be making process decisions.

A key point to be made early in this chapter is the assumption that a hierarchy of data and information exists, as shown in Figure 9–1. This hierarchy was first presented by Mr. Dan Appleton to depict how information actually evolves.[1] Notice in this hierarchy that "data" really have no value, until they are merged with the "context" for the data to form "information." For example, a stream of digits representing a data stream might be 12345. But until the context is pronounced to be ZIP code, one does not know that the stream is actually a location within the United States, specifically Schenectady, New York. "Information" is then merged with background and experience to form "knowledge." This transition was highlighted in Chapter 5 where the limitations of generative process planners was outlined. Background and experience are very hard to capture in a knowledge base, hence the difficulty in

**FIGURE 9–1**
**The Pyramid of Knowledge**

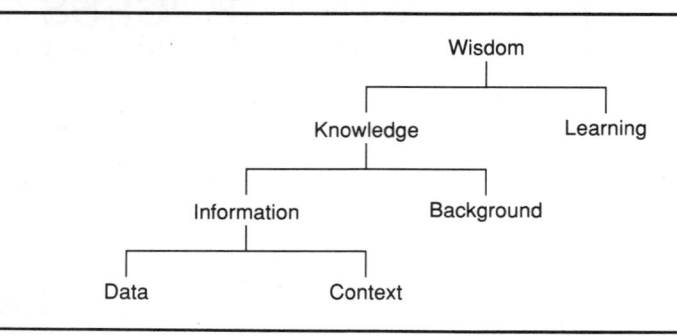

```
                    Wisdom
                      |
          ┌───────────┴───────────┐
       Knowledge                Learning
          |
    ┌─────┴─────┐
Information   Background
    |
 ┌──┴──┐
Data  Context
```

using the advanced planning tools for modern applications. Ultimately, the top of the pyramid is "wisdom." This is the total amalgamation of all learned from below it.

Process design must be integrated with product design. This element is critical for future systems success. Design systems and manufacturing systems must be integrated using consistent definitions of product and process and downloading and uploading information from sources to sinks that are very efficient and operate in a proactive, closed-loop system.

To support this, the notion of computer integrated planning (CIP) is developed in this chapter. Cost, time, and quality dictate these system requirements. As outlined in this text, integrated process design and development systems are required to:

- Be closed loop.
- Support flexibility and efficiency at the same time.
- Allow part family planning with setup times approaching zero.
- Continuously improve in the execution of each of these processes.
- Have flexible planning systems support.

This system requirements list provides the basis for the integration of the planning systems, the control systems, and the execution systems as they relate to the manufacturing enterprise. Both feedforward and feedback must be present from execution to control and control to planning. This closed loop must be consistent with the timing elements necessary for decision support. The integration of process planning and scheduling is one of the integral parts of the integrated planning system. Often these two tend to be inflexible and unintegrated; but we must integrate. CAD data sets, process requirements, equipment capabilities, supplier-based capabilities, master schedule needs, time lines, and so on must all be provided to the process design team as efficiently and effectively as possible. The result is a real-time creation of the best routings given up-to-the-minute shop status.

Certain prerequisites for integration are needed to support this. For example, we must have consistent product and process definitions. We must have true equipment characterizations and a clear understanding of their capabilities. We must have tooling and fixturing that allows quick change and quick changeovers and yet apply rapid and consistent and total quality to the process. We must have the ability to interpret the master schedule by part family, and we must have the flexibility to adjust routings based on current status of needs and capabilities between the customer and the factory floor.

To create this, we have developed the following steps for integration. First, create process requirements and alternate routings. Second, create equipment capability files for the factory and of other suppliers. Third, interpret the master schedule and create shop schedules based on this master schedule, consistent with customer needs and shop capabilities. Fourth, merge those requirements with the capabilities to create the best routings for that particular moment. And fifth, maintain ongoing flexibility to allow for material, equipment, and/or tooling failures and be able to dynamically reschedule the floor for any occurrence.

The notion of *computer integrated process* hinges on the ability to integrate within and between the enterprise functional areas. This chapter will develop two fundamental looks at the way information must be captured, stored, and retrieved. The chapter will look at process design information as it relates to the total quality management ability to improve the process. Next, a "strawman" hierarchic architecture for integrated process design and development is presented to discuss connectivity. The third section in this chapter introduces computer integrated planning. Then the process control information system needed to support CIP is introduced, and the fifth section deals with the integration of planning with the control information system. The sixth section introduces the notion of integrated shop floor control, and the chapter ends with an overview of the need for financial system integration, legacy systems management, and proactive management to support these very aggressive ideas.

## SECTION 9.1 PROCESS DESIGN AND DEVELOPMENT INFORMATION SYSTEMS—AN OVERVIEW

Planning in tomorrow's factory must represent the flexibility necessary to support an economic order quantity of one and yet have the rigor to withstand and support 100 percent shop floor data integrity. Today's planning must be done six to nine months before making some parts, but when the parts ultimately are made, the shop is changed. Results from a National Science Foundation study recently found that over 50 percent of the process plans have at least one step in the plan modified during the actual fabrication process.[2] To resolve this, two approaches can be taken. One is to continue to attempt to add rigor to the existing system as we use it today. A second is to develop a strategy to migrate to a new structure or a new architecture for future planning systems.

In this text, we stress that the modification and continuous improvement of existing systems are fundamental to the success of any organization. But this chapter later presents a new approach that addresses an architecture that can accommodate very small order quantities, reflect advancements in technology, and also serve as a consistent, rigorous planning tool that integrates technical systems with business systems.

Finally, technology gaps and cost-accounting impacts will be addressed in this overview. Traditionally, design engineering has been the driving force behind product manufacturing. Designers often specify a product that functionally meets the needs of the organization with little industrial engineering or manufacturing engineering support. As was shown in Sections 1.3 and 5.3, the leveraging opportunities for integrated process design and development occur early in the product design phase. Traditionally, the process engineer has not been an active participant in this phase for two reasons:

- It was not perceived to be the process engineer's domain.
- Little quantitative data were available to make process decisions while only dealing with a sketch of the product.

These process engineers have constantly been faced with the problem of how to manufacture the item at minimum cost and/or time and yet live within the confines of existing equipment, production loadings, and fluctuating management objectives. Today, the integration of the design systems and the manufacturing systems in a concurrent mode provide a major opportunity to significantly reduce this problem. Working together early in the concurrent engineering phase would allow process engineers to participate actively in the product design to ensure that manufacturing capabilities are present.

However, at some time in the course of the product, the product must be created. Integration, from design through product manufacture, is conceptually possible and, more important, economically desirable. But we ultimately have to make this product, given the finite capacity of each of our facilities. Since direct labor currently accounts for only 10 to 25 percent of the cost of manufacture, it is no longer possible to attempt to significantly improve productivity through direct labor savings only. In fact, it has been stated, "The major opportunity for improving productivity is in organizing, scheduling, and managing the total manufacturing enterprise."[3]

Design and manufacturing system applications are being implemented at an increasing rate. Yet very little success has been achieved through the integration of CAD and CAM in some of these installations because of the lack of consistency (rigor), the lack of metrics, and the inability of the factory floor to be adequately reflected in the transition from the planning to the control to the execution phases. Many of the barriers between design and manufacturing exist as they have since the beginning of the Industrial Revolution.

In the current manufacturing environment, a new part is designed using a computer-aided design system. Most CAD systems today use "stick figure" representations of the product that force the user to interpret the drawing. The product is defined by the lines that make up the outside envelope of the product and the lines that define the internal shaping of the product. By sticking these lines together, it is possible to *interpret* where the product is and where there is only air, but this interpretation is subject to error and requires experience. Advancements into features-based, solid modeling systems will help in the visualization of the product and will also help the planning systems designers by allowing information to be stored in the CAD data file. In features-based systems, the product is designed using a series of primitive shapes that can be easily interpreted. By either adding or subtracting these shapes from the initial form of the product, it is easily seen what the part is and where the air is. It is also much easier to interpret the shape though computer analysis of the features to create the process plans.

On approval, the design is released to manufacturing. The manufacturing engineer can then obtain hard-copy drawings of the part or view the part using a graphics computer terminal. Either way, using the information in the CAD system, the manufacturing engineer must design the manufacturing process.

In most manufacturing organizations, communication between design and manufacturing engineering is becoming more evident, yet rigorous models of the overall design-manufacturing process using the TQM notions are lacking in many cases. The manufacturing engineer uses techniques such as group technology in a variant process planning mode to minimize the design of the manufacturing process. However, someone must manually view the part design in the CAD system to code the parts and then interact with the process planning system. So in a current manufacturing environment, there is a barrier between design and manufacturing functions. One significant bridge across this barrier is the development of computer software systems to recognize part features in a design data base followed by the development of a generative computer-driven system to design the process plan.

In most manufacturing environments, it would be naive to think the process plan developed at this point would describe exactly how a part would be produced. The plan probably would describe the ideal way to produce the part given the machines and tooling available to a company. However, when the time comes to perform specific operations, the scheduler may find that specific machines and/or tools are not available. Consequently, alternate routings must be considered and developed. Now, because of time considerations, decisions as to how to accomplish a specific operation, and future dependent operations, must often be made with very little information. As a result, the actual processes a part goes through may be very different from the original plan designed by the manufacturing engineer.

## SECTION 9.2 AN ARCHITECTURE FOR INTEGRATED PROCESS DESIGN AND DEVELOPMENT

To introduce an integrated process design and development system, an architecture of systems is shown in Figure 9-2. This is simply a sample on which to build. Every company has varying systems architectures. The one shown in Figure 9-2 was developed by a major American aerospace company with funding from the U.S. Air Force's integrated computer-aided manufacturing program (ICAM). The results of this study demonstrate how an architecture can be created to support the integration needed.[4]

The results of the study show that the move from shop floor modernization to "above the shop floor" integration has presented quantum leaps in potential productivity improvement. With an industrywide standard cost system, the aerospace company found that the real cost of the product was driven by manufacturing support costs and material costs. Hence, improving efficiency and effectiveness and developing integrated systems presented major opportunities for productivity gains that would result in significant cost reductions.

By launching into the ICAM program's project priority 1105 that addressed "Conceptual Design for Computer Integrated Manufacturing," the company went a step further and outlined a framework for the generic "factory of the future." The project addressed the following targets for success in the aerospace marketplace in 1995 (these are not limited to aerospace):

1. New, evolving organizational structures based on alignment of functional activities.
2. Information management as a critical resource.

**FIGURE 9-2
A Generic Factory of the Future Concept**

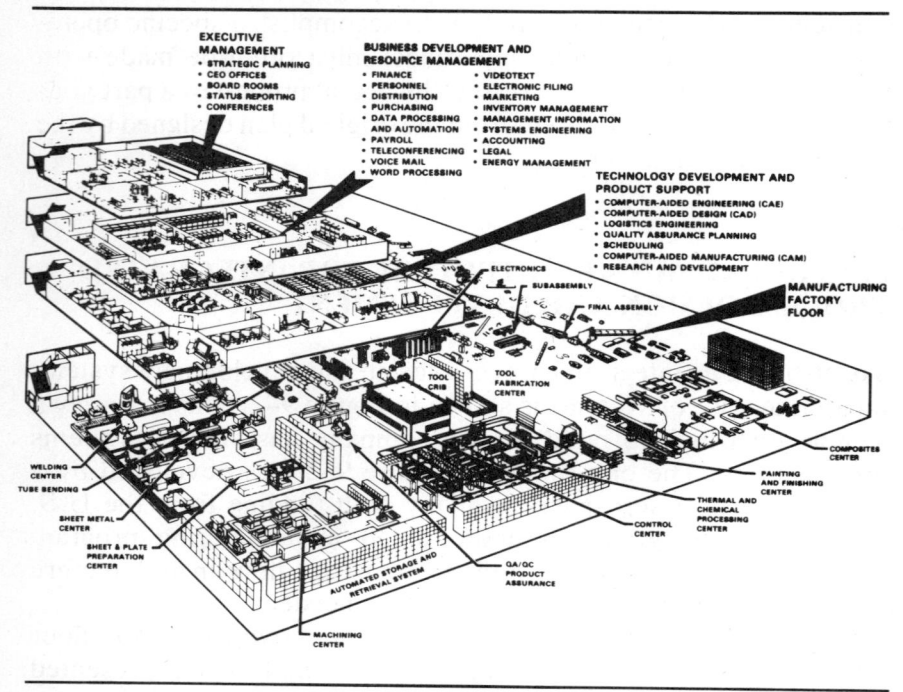

3. Strategic leveling of manufacturing technologies to ensure competitiveness.
4. Maximum utilization of talent/expertise.
5. Standardization of the design process.
6. Increased emphasis on quality assurance.
7. Flexible processing of materials in a continuous flow.
8. Maximized, long-term return on investment (ROI).

The project produced a functional hierarchy as shown in Figure 9-3. This chart depicts a hierarchic structure of organization that is designed to produce a continuum of information and work flow. The result of this effort was a structured road map with the objective of

## FIGURE 9-3
## A Functionally Hierarchic Architecture

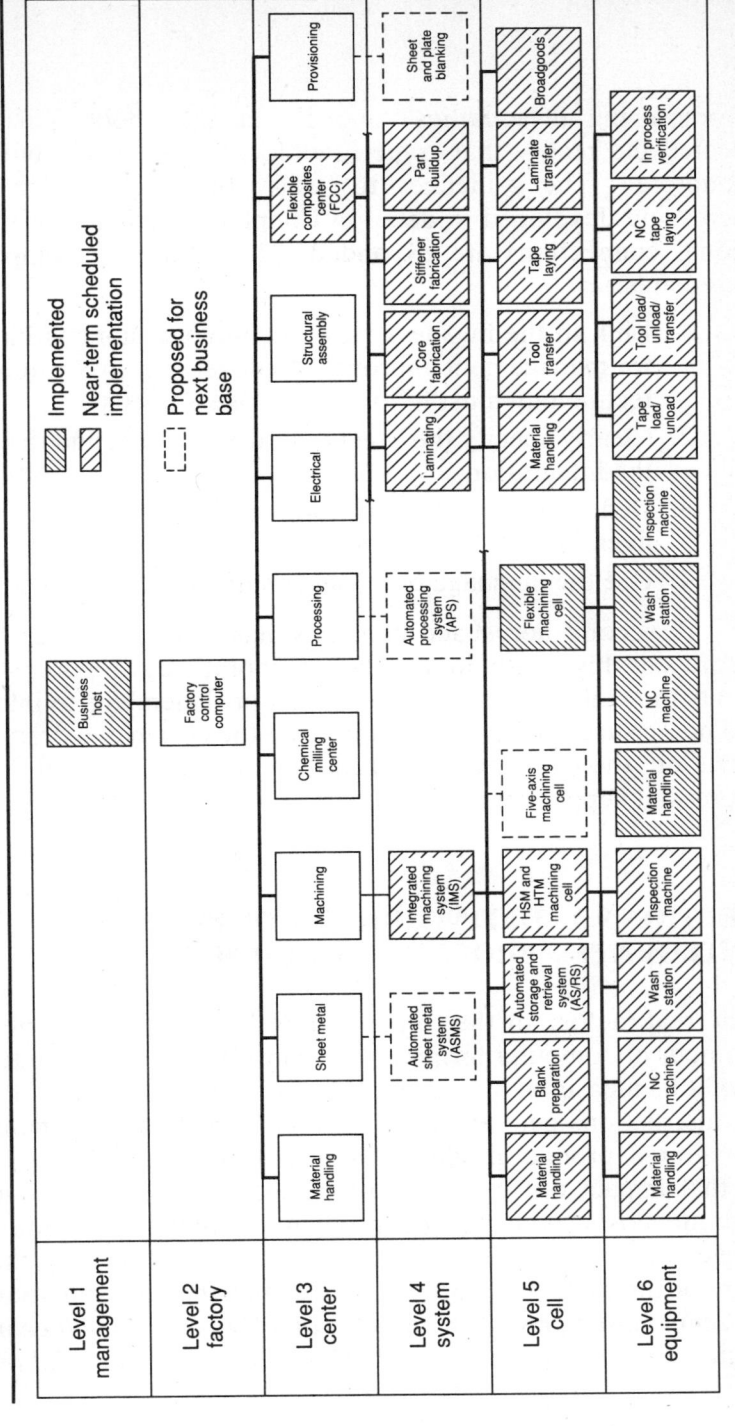

Source: Reprint courtesy of the Society of Manufacturing Engineers.

modernizing all activities "above" the shop floor. This also reflected the reality that some roadblocks to change would be encountered. The guidelines to "sell" this concept are outlined here, and developed in detail in Chapter 12, where the interfunctional connections are expanded. The aerospace company recommends:

1. Link to the business base; technology alone is inadequate for management approval. As developed in Section 1.4, the technology must follow the business needs.
2. Clearly specify the "facts" of the business and the need for the integration.
3. Define a clear, proactive project plan with intermediate deliverables and milestones.
4. Use the language of management to sell the project.

A second aerospace company has drawn another figure to represent this notion in a slightly different way. Figure 9–4 represents the real need for three organizational elements to integrate to compete successfully. These three are the design and drafting organization, the planning and scheduling organization, and the operations and support organization.

## SECTION 9.3 DEFINITION OF COMPUTER INTEGRATED PROCESS REQUIREMENTS

The real factory of the future cannot be measured simply by the number of computer terminals or reams of computer printout. Nor can it be measured by the number of computer hookups or millions of instructions per second or number of personal computers. The integrated product-process that will work must be measured in actual impact on two key performance characteristics—reduced product costs and increased throughput—that directly add value to the customer and make the customer successful. Hence, the definition of the factory of tomorrow becomes the amalgamation of *100 percent good parts* produced very *quickly* in very *small economical batch sizes* at *low cost* supported by only a *few indirect laborers* that

**FIGURE 9–4**
**The Integration of Three Organizations**

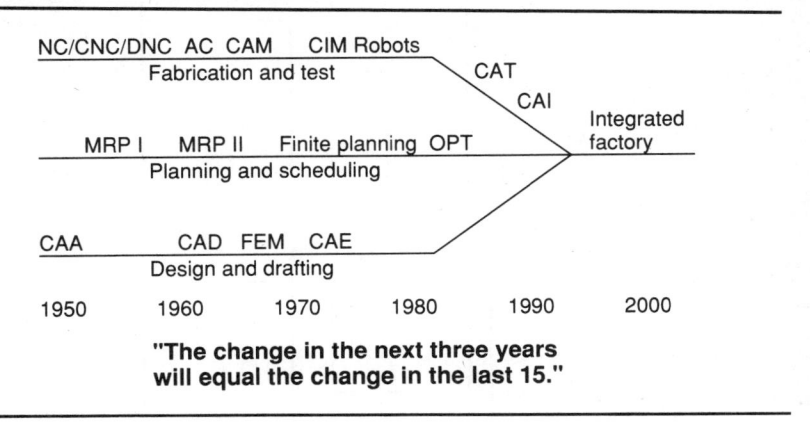

```
NC/CNC/DNC  AC  CAM     CIM  Robots
       Fabrication and test        CAT
                                      CAI
                                            Integrated
   MRP I   MRP II   Finite planning  OPT    factory
        Planning and scheduling

CAA          CAD  FEM  CAE
        Design and drafting
1950      1960      1970      1980      1990      2000
```

"The change in the next three years
will equal the change in the last 15."

directly *meet the requirements of the customer*. Ramifications of this definition transcend through each system design. The requirements for real integrated product-process systems can be summarized as:

1. Closed-loop operations for process control where nonperformance operating regimes are well understood, documented and known to all.
2. Transfer line efficiency with job shop flexibility and optimum utilization of bottleneck operations.
3. Part family planning resulting in family tooling allowing setup times to approach zero.
4. Totally integrated, continuous processing, real time tracking and control systems with the ability to effectively modify the allocation of components to assembly based on a real time feedback system that sensed the shop floor situations.
5. Flexible planning systems that allow perturbations in requirements or availabilities to be reflected in process plans and routings.

**FIGURE 9–5**
**Flexible Control System Architecture**

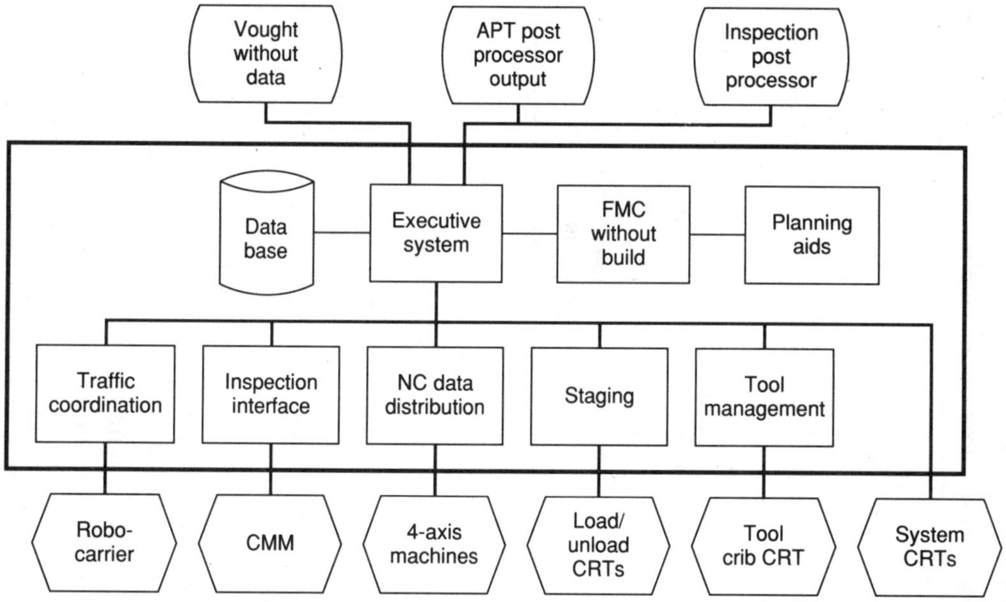

Source: Reprint courtesy of the Society of Manufacturing Engineers.

## SECTION 9.4 PROCESS CONTROL INFORMATION SYSTEMS

To address these requirements, integrated systems must be created, and integration must occur at all three levels of systems—information integration, controls integration, and material flow and tooling integration.

With respect to information integration, we must achieve the planning, tracking, and control necessary to provide information that is available where needed in the form needed. This often requires various pieces of computing hardware to peruse various data bases all with absolute data integrity.

Controls integration establishes the requirement to achieve the closed-loop process control that can link each processor and operate each process. Control strategies and systems must be integrated. This forces the control system to know what parameter set is necessary to make a good part and to monitor these parameters to ensure a good part is made. It also forces the control system to know where the parts are at all times. Figure 9–5 presents another "strawman" architecture for this integrated process control system.[5] By establishing a network that flexibly links all functions in the cell, response times can be minimized.

Material flow integration is also represented in Figure 9–5. The objective is to achieve transfer line speed with job shop flexibility, know where parts are at all times within their lifetimes, and have them handled in an efficient manner such that we maintain their orientation with known moves.

When these three integrations are achieved, the integrated process design and development system can occur. Incumbent in the investment strategy to support this system is the recognition of where the most productive processes are needed. Based on the theory of constraints espoused by Eli Goldratt,[6] key bottleneck operations must be the focus for process improvement to produce the best product in the least time.

## SECTION 9.5  DESIGN OF THE INTEGRATED MANAGEMENT AND CONTROL INFORMATION SYSTEM

Having defined the target, the next step is to define the integrated systems to meet the target. Planning and scheduling play a supreme role in proper utilization of resources. Today, two inflexible systems often exist. One is the MRP-based business system that builds on bills of material and determines how many of which parts are needed, how these parts are needed, and how these parts shall come together. The second is the process planning-based technical sys-

**FIGURE 9–6
A Computer Integrated Processing System Architecture**

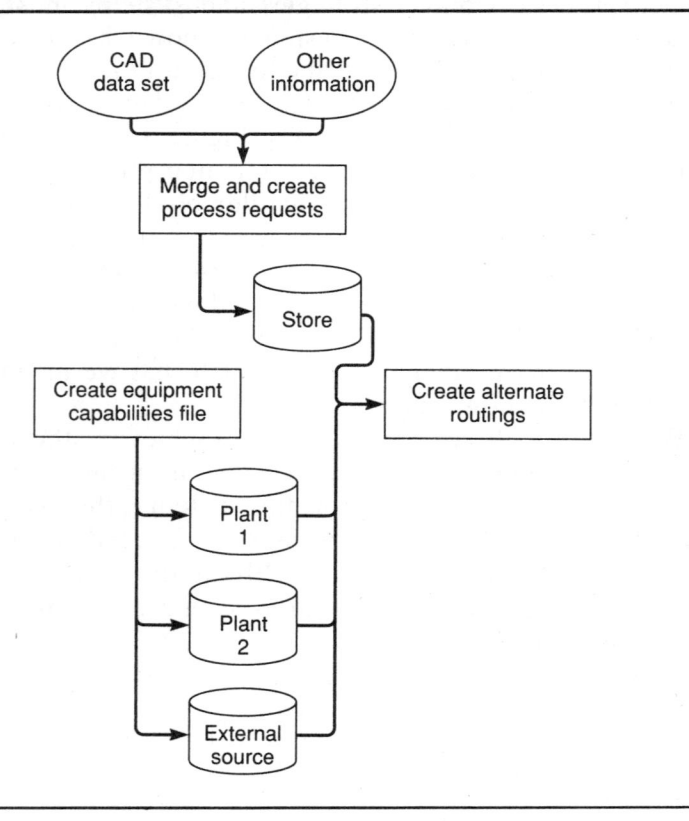

tem that reviews the inherent characteristics of each part and conceives the means by which existing equipment can produce the part.

But in the integrated process system, these systems must reflect the ever changing environment of marketing, sales, procurement, maintenance, suppliers, and so on to create a flexible system. In this system, bill-of-material type business systems must merge with part-type technical systems in an on-time atmosphere to create requirements and routings that optimize cost and time. By necessity, the computer integrated processing, or CIP, system must know the processing requirements of each part and also know the capabilities of each piece of equipment to perform that process.

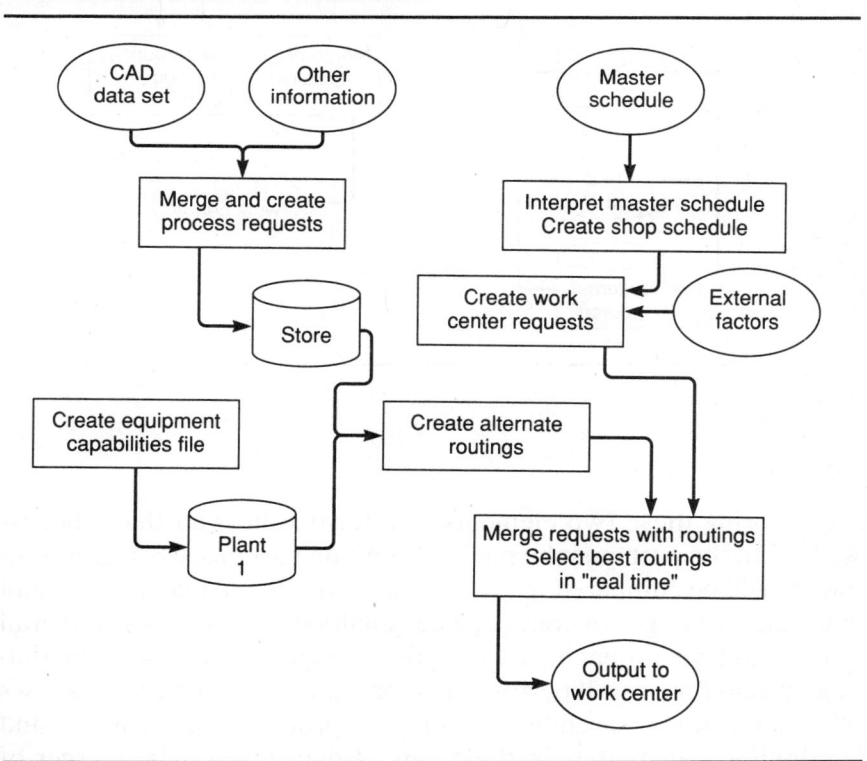

**FIGURE 9–7**
**An Extension of the CIP Architecture**

**FIGURE 9–8**
**Finite Capacity Extensions of the CIP Architecture**

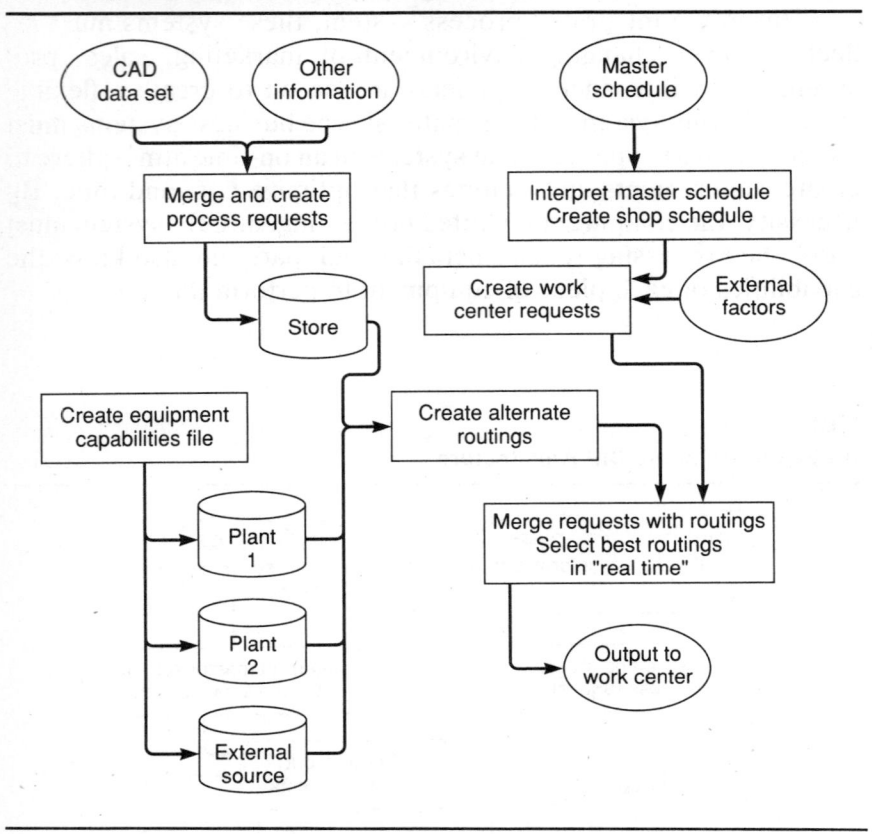

Storing these two elements of information is all that's necessary until the part must be made. Then, the part processing requirements will be compared to the available equipment capabilities, and alternate routings are created. The penalty factor in terms of overall costs and time for each part shall then be quantified to the scheduling system for overall performance optimization. Figure 9–6 shows this data storage architecture of part process requirements and equipment capabilities in the forms of data bases. The merger of requirement with capability creates the alternate plans that will be needed.

But this computer integrated processing system does not support just process planning. This system should be used in every step of the planning process. For example, in each make-buy decision, the planning system can compute the cost and time to make a part with existing equipment in one plant versus outsourcing to a supplier. A second straightforward application would be in capital planning for the most-effective equipment addition and deletion to the current capabilities.

Finally, costs and times for any equipment configuration can be assessed rigorously and quickly. This can be seen as an extension of the basic planning system structure shown in Figure 9-7.

Finally, a schematic of the finite capacity system using alternate routings of the computer integrated processes is shown in Figure 9-8. This allows flexible scheduling until the actual time of need. All flexible manufacturing system cells in the future will need this capability.

## SECTION 9.6 THE DEFINITION OF MODERN, INTEGRATED PROCESS EXECUTION INFORMATION REQUIREMENTS

The history of such areas as shop floor data collection, factory management and control systems, and new cost-accounting integration is well established and can be traced to developments by Joseph Orlicky, George Plossl, and Oliver Wight, in their fundamental looks at new methods of operations management. The development of the shop floor control module in such systems as manufacturing resource planning (MRP II) marks the beginning of the recognition that information must be integrated to achieve overall integrated process design.

The notion being presented here is that the information integration will probably be on a computer system, and it will be called a factory management and control system (FMCS). In some cases, the functionality of shop floor control and factory management and control systems have been enhanced to such a degree that a change in terminology is needed to differentiate between traditional shop floor control and factory management functionality and the new and improved factory management and control systems. Many labels have been pegged to FMCS applications, including such titles as:

- Factory control systems.
- Shop floor control systems.
- Factory management systems.

The lack of a consistent title is partially due to the lack of consistency between functional elements within the system. This confusion has led to the different categorizations of these systems.

Statistical process control and other elements of the total quality management and geometry definition thrust for an integrated process design must be included as inputs to the FMCS. Traditional shop floor control systems have the same requirements regarding the level of detail. To take full advantage of the integrated process design system, the islands of functionality defined in this text must be integrated. To do so, we must understand the required information elements and how these information elements can be linked. Factory device sensors and controls must be linked with information systems and with planning systems to fully integrate the overall process.

Configuration management is another shortcoming of the overall factory management and control system. In today's vocabulary, the MRP II system provides an as-planned manufacturing process instruction set. However, it becomes very difficult, almost impossible, to monitor the changes that occur during the shop floor activities of a product unless a fully integrated system is in place. And given the results presented earlier, 50 percent of the process plans may be changed once they reach the factory floor.

We must provide the ability and the capability to control changes and adjust to all changes within the factory management system as changes occur on the factory floor. The as-built configuration must be tracked and recorded in a consistent and integrated fashion for us to track that product as it enters the field.

Problem identification also encompasses many of the shop floor and data-collection deficiencies stated earlier. The lack of visibility of quality problems, production problems, and reporting status to interested parties is a serious need that must be resolved in an integrated process design system. To manage this CIP system, almost real-time visibility is needed.

"Real time" is a specific terminology related to controlling certain processes on a microsecond-to-microsecond basis because changes in the process might occur very rapidly. However, the terminology used here is that of "on time," where on time defines the ability to respond and have information available to a decision maker in a timely manner. In many cases, this need not be real-time data collection and data management. On time might be minute-to-minute, hour-to-hour, or day-to-day updates. We are finding that the migration to a time domain allows us to understand the timeliness requirements for when decisions will be made, how they will be made, and what information is going to be required to make those decisions.

Greene defines factory management system as a "system which provides dynamic on-line monitoring and control of the manufacturing shop floor resources and processes."[7] Greene's title and definition taken together accurately identify the two key elements of these systems—management of resources and control of processes. Factory management and control systems are the integrated process design support tool that allows us to effectively execute the overall process on the factory floor. Factory management and control systems have not yet reached maturity. A 1988 survey showed that factory management and control systems had penetrated only 7 percent of the potential marketplace.[8] The concept of factory management and control must mature to encourage system development and increase industry acceptance. This will happen as we develop the knowledge contained in this text.

MRP I and MRP II are both examples of defined bodies of knowledge that include logical, coherent, and well-documented streams of thought. Both are taught in schools and widely understood. The concept of integrated process design using factory management and control systems as an element to perform process execution is not yet widely understood by the market and, in some cases, the schools.

Figure 9–9 shows the classic factory management and control activities defined in current literature. This exercise focuses on those elements within the major shop floor control area that deal with order release, detailed assignment, performing fundamental

**FIGURE 9–9**
**Classic Factory Management and Control**

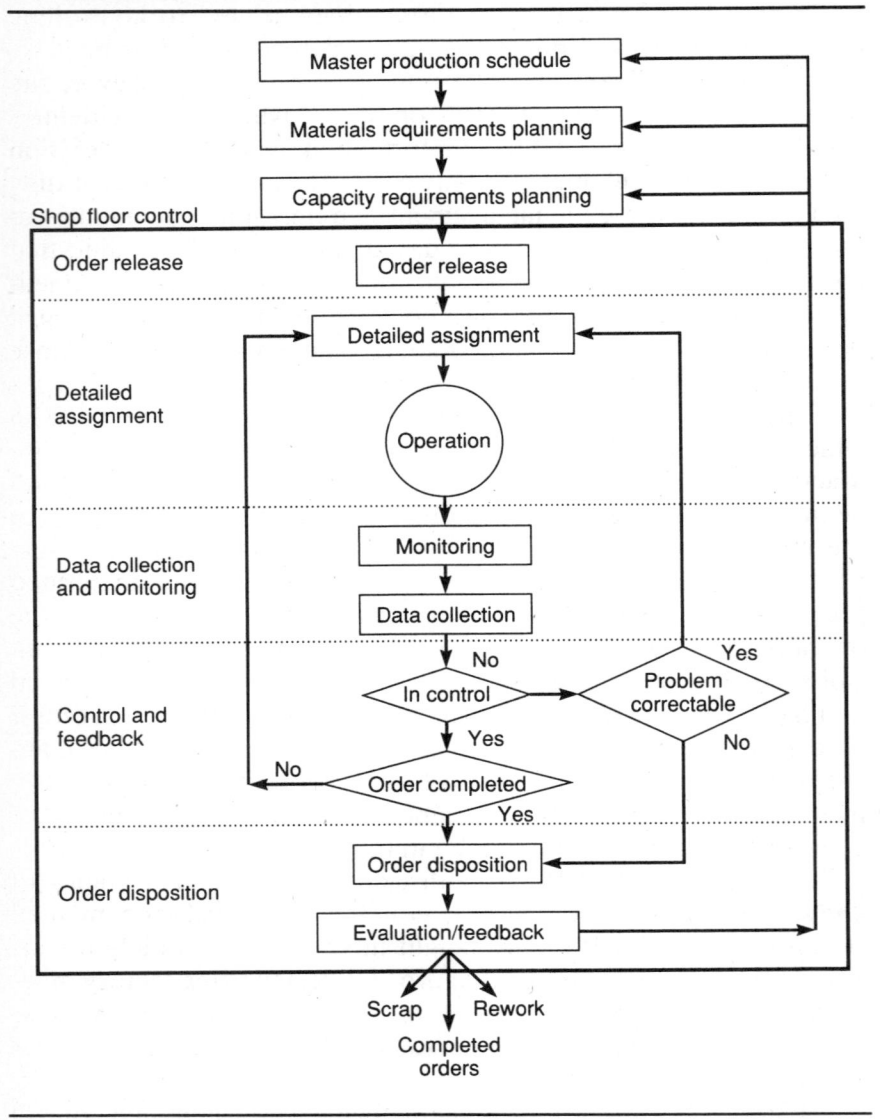

operations, monitoring data collection, testing to see whether the process is in control or needs to be fixed and corrected, whether the order has been completed, or whether another cycle must be performed, and ultimately, the order disposition, the evaluation and the feedback, and then the disposition of the finished goods. Detailed assignment involves scheduling the shop by matching factory resources, which, in this case, are work centers with the released work order routings. This includes work order sequences by work order priority.

Typically, priority is determined by earliest due date. As a result of priority planning, a dispatch list is generated that contains all the work orders in priority sequence waiting to be processed at a given work center. Data collection and monitoring links the planning system and the execution system. Work-in-process tracking is performed to identify the progress of the work order as it moves through the factory. Control and feedback involve corrective action when work orders deviate from the original plan due to rework, scrap, or some other element of the integrated process that is not performed successfully. Corrective action can include making short-term adjustments in work rates, overtime, lot splitting, subcontracting, and so on.

The final activity in factory management and control involves order disposition. This includes transferring responsibility out of the factory management and control system environment when an order is completed or scrapped. At this time, the order is no longer part of the factory management and control system, and the appropriate inventory data bases are updated.

Factory management and control systems must complete proper priority planning if they are operating in an integrated environment. This priority planning is the logic that determines the sequence of jobs to be processed at each work center in the factory. Factory management and control systems consider typically only the current status of a job, that is, on schedule, ahead of schedule, or behind schedule. But as we migrate to the integrated process design environment, many, many issues must be looked at in either a look-ahead or a look-behind environment. Many factors must be considered when producing this overall integrated system, such as status of later workstations, current workstation setup, equipment status, tooling status, required operator skill levels, and operator availability. As we discussed earlier in this text, if we look at tooling and fixturing in a single-minute-exchange-of-dies scenario, then all these issues become fundamental to the opportunities to create an economic order quantity of one with zero setup time between lots.

The next level of sophistication for factory management and control systems is to elevate the capabilities to support complex priority planning logic needing visibility, looking ahead and behind to consider all of these factors.

In addition, the factory management and control systems must integrate with the overall quality management functionality. The ability to define quality parameters and link them to routing operations and machine tests is necessary to achieve integration. Statistical process control in a timely manner is needed to detect trends and recommend actions to preclude producing any bad parts. There are also interesting trends to link the statistical process control results to downstream operations by lot to ensure that all of the processes on the lot build on the previous processes in proper fashion through real-time compensation.

Finally, the process plans developed and introduced in Chapter 5 must be fully documented and complete for the integrated process design to develop a proper monitoring and control system. The production instructions, or process plans, must be detailed to the level such that a shop floor operator knows exactly what to do in performing each of the routing operations. Factory management and control systems provide only textual information at this time. But in the integrated directions of future systems, as we will discuss in Chapter 10, graphics and text, even audiovisual and multimedia representations of requirements, will be necessary to provide shop floor operators with all the data necessary to perform the task adequately.

In an assessment of the existing state of the art of factory management and control systems, the following summary of all the information elements found in systems today is presented. A rigorous survey conducted at Arizona State University identified elements of information needed for factory management and control systems.[9]

1. Information about the factory employee.
2. Information about the operation.

3. Information about the product.
4. Information about the process plan.
5. Information about the work unit.
6. Information about the schedule.
7. Information about the workstation and the capabilities of that workstation.
8. Information about the resources.
9. Information about exceptions that occur during the course of the actions.
10. Information on each of the actions.
11. Alarms that must be captured and noted to each of the factory employees when actions occur that are outside of acceptable limits.

In Figure 9–10, each element of information is shown with the various properties and services that must be performed to develop the detail to meet the process execution information requirements.

Finally, as we begin to assess the metrics of the 21st century as shown in Chapter 7, and recognize that time, quality, cost, and responsiveness to changing customer demands are necessary, we must begin to develop an assessment process of the process execution information data elements that allows us to migrate to the cost-accounting system of the 21st century, which is typically defined today as being activity based.[10]

If we look at each of the elements of information shown in Figure 9–9 and Figure 9–10, as representing the overall functions performed for any product as it progresses through an integrated process execution function, then each of these functional elements must develop their own cost-accounting integration capabilities using activity-based costing as the guideline.

## SECTION 9.7   REQUIREMENTS FOR FINANCIAL SYSTEMS INTEGRATION

The typical cost of goods sold (CGS) for a $100 million sales company that produces a sophisticated or semisophisticated product has changed dramatically over the past 25 years. In the 1960s and

**FIGURE 9–10**
**Object-Oriented FMCS Details**

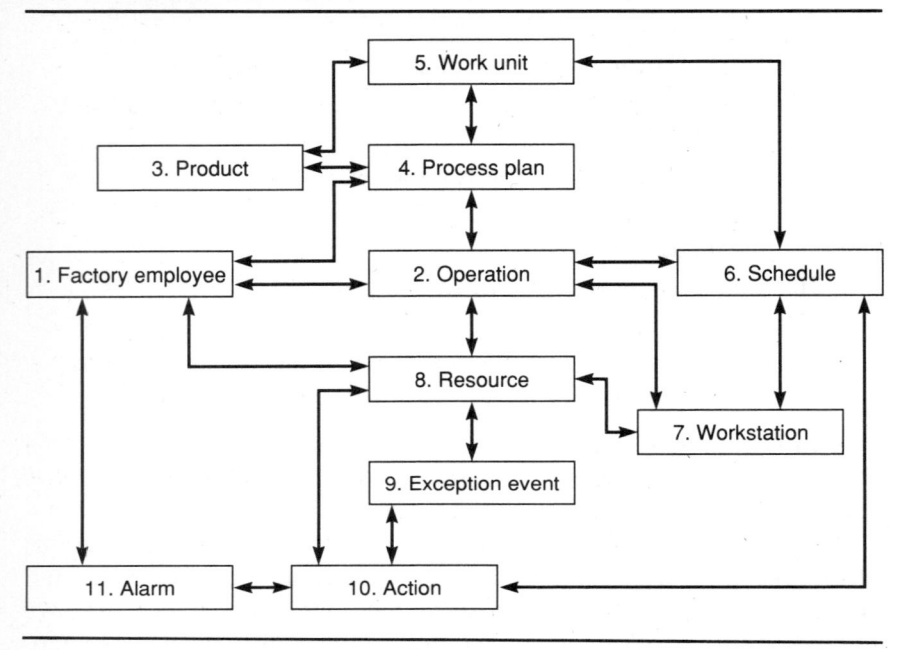

[Diagram showing relationships between: 1. Factory employee, 2. Operation, 3. Product, 4. Process plan, 5. Work unit, 6. Schedule, 7. Workstation, 8. Resource, 9. Exception event, 10. Action, 11. Alarm]

1970s, a major portion of the CGS was direct labor, and cost accounting used the direct labor content as the basis for calculations. But as the direct labor content has been reduced dramatically, the use of direct labor as the basis has become a major debate. With direct labor content at 5 to 10 percent of CGS, companies find this is not a good predictor of the total CGS. This has led to changes in cost-accounting methodologies, such as the trend toward activity based costing. Mr. Robert Kaplan stated this need as follows:

> As competition shifted to low-cost production of commodity products, the company had to develop a new cost system to give unit managers more reliable information about their production efficiency.[10]

The reason so many companies are having such difficulties can be directed to the three different functions this system must perform:

**TABLE 9-1**
**Different Views of Cost Data**

| Functions | Frequency | Degree of Allocation | Nature of Variability | Degree of Objectivity |
|---|---|---|---|---|
| Inventory valuation | Monthly or quarterly | Aggregate | Irrelevant | High |
| Operational control | Daily | None | Short-term variable and fixed | High |
| Product cost measurement | Annually | Extensive | All variable | Low |

1. Provide accurate inventory valuation for financial and tax statements.
2. Provide operational control with real-time and on-time feedback to the operations management team.
3. Provide accurate individual product cost measures.

The 21st century company can no longer afford cost systems that provide only the financial reporting data and neglect the timely support of operations and product managers. By integrating multiple looks at the cost data, companies are beginning to extract the correct data for the correct use. As shown in Table 9-1, Kaplan pinpoints the different functions that have different demands.

Analysis of this table shows that the three categories have significant integration demands. Inventory valuation often divides these costs—labor, materials, purchases, and factory overhead—between items sold and items still in inventory. But inventory valuation continues to often use an allocation formula based on direct labor content. However, operational control requires much greater visibility in areas such as:

1. The frequency of the reported information directly linked to the cycle of the production process.
2. The fluctuations in costs and how they should relate to fixed versus short-term variations.
3. The true allocation of costs based on the real departmental involvement.

4. The integration of nonfinancial measures such as yield, defects, output, setup and throughput times, and physical inventory levels.

Finally, product cost measurement adds new dimensions to the integration requirements for the integrated product-process environment. Product cost needs aggregated data on issues such as:

1. Actual allocations of costs from various participating departments.
2. True allocation of longer-term variable costs and the mixture of these over time with the fixed costs anticipated to give a very clear picture of the real product cost over time.
3. Access to the total enterprise costs often added to the product cost, such as distribution costs, selling costs, and advertising, to name a few. Research and development (R&D) costs must be handled in a very systematic manner to capture the true value they added to this product versus their value to future products.

The implications of the need for tremendous visibility into total system costs plus the timeliness issues for unit managers to senior executives force major integration issues into the factory management and control system and how this system integrates with planning and other control functions.

## SECTION 9.8 LEGACY SYSTEMS AND THE NEED FOR DYNAMICALLY MIGRATABLE SYSTEMS

Many companies find that their information systems cannot economically be maintained at the most effective configuration because of the "legacy" of hardware and software that have been installed. Difficulties in supporting a typical, large software system through its life cycle cannot be overlooked in the migration to integrated process design and development. An ideal solution would be the migration to dynamically migratable systems (DMSs). A DMS is an architecture for a system that protects the previous investment while allowing the upgrades needed for modern performance.

These DMSs are expected to maintain the look and feel for the users for consistency and to minimize the time to acceptance by the users. This demands a new kind of built-in flexibility on the part of the systems designers. System migration will be generally easier if systems can interoperate with other systems in a seamless way and if they are built on the foundation of portable operating systems.[11] The ease with which software can be adapted is a critical requirement.

Four major issues exist for this DMS approach to managing legacy systems:

1. Coupling between data structures and procedures.
2. Casual software documentation.
3. User-developed software desires.
4. Software maintenance.

Two paths are offered to address these issues of DMS. The first is to deal with the dynamic migratability issues directly with new and novel approaches to systems design, and the second is to deal with the current approach to systems design and attempt to minimize the anticipated maintenance costs.

The logical steps to develop new and novel approaches begin with the following:

1. Revamping of system architectures.
2. Encapsulating data and procedures.
3. Attempting to hide the data abstraction.
4. Clarifying the logic that goes into the system and converting that to be aligned with user perception.
5. Developing self-documenting systems.
6. Developing consistent iconic interfaces.
7. Reducing the bulk of software code.
8. Building in the reusability of software.
9. Building in the modularity and modifiability into software.

Many of these steps hint at the use of a new technological approach for system definition called "object-oriented programming" (OOP). The OOP philosophy is promising but not yet proved. The cornerstone of the approach is that data and procedures related to the entities of the system are represented in a structure called an "object" and the data are private to the object and can be accessed only by using the procedures contained in the object.

In summary, the migration to DMS requires a new paradigm for system definition and system development. New tools such as OOP show significant promise in allowing new, migratable, well-documented systems to be created in such a fashion that cost-effective migration is feasible. But not all is proven with the OOP approach. Slow execution and high memory usage often arise in OOP discussions.

In an exercise conducted to demonstrate an integrated systems approach by the U.S. Air Force in 1987, the users were faced with the following question: "A unique tool has broken on the factory floor and it will take some time to replace it. What orders are affected and how should each be handled?" In the industry demonstration of a system called the "integrated information support system" (IISS), two computers with two different data structures were integrated. First, the unique tool had to be tracked back to the process plans where it was referenced, and then the process plans were tracked back to the unique parts affected. This was conducted on a Digital Equipment computer. Once the part numbers affected were known, the data were transferred to an IBM computer that used the part number list to trace back to the orders. The time to perform this in an aerospace company in 1987 was measured in hours, maybe in days. Yet the demonstration was concluded in minutes. The systems developed were slow and inefficient from a computer usage view, but from a user view, an answer in minutes made the decision-making task feasible. We must continually remember that a total systems approach is mandatory for integrated process design and development.

## SECTION 9.9  PROACTIVE SYSTEMS DESIGNED FOR PROACTIVE MANAGEMENT

If such a computer integrated process system is so powerful and so necessary for information integration, then why does not one exist? A combination of factors are involved.

First, the segregation of process requirements from equipment capabilities is quite new. Normally, the route sheet assigns specific machines to specific tasks months in advance of the actual creation of a part. This is why over 50 percent of the routings are often not followed. But with the advent of new technologies, such as object-oriented programming and others, the segregation of different object characteristics becomes now technologically feasible.

Second, the group technology structure for parts, processes, and equipment does not exist in such a manner that the relationships of requirements to capabilities can easily be noted. Here again the advent of modern technologies allows this technological advancement to now be feasible.

And third, the relational data structure necessary for efficient handling of transactions is now just becoming available. A quick review of each of these points shows that none is insurmountable, but also that no system as described has been built.

In today's planning environment, the process plans serve as the basis for predicting job costs and routing of jobs through the shop. Alternate routings cause havoc in both systems, but if we view the future factory the way Bob Kaplan does, direct labor is probably going to approach a very small percentage of the cost of goods sold. Variable costs of today will probably approach fixed costs of tomorrow and vice versa. For this view, the indirect costs to plan will undoubtedly greatly outweigh the direct costs to accomplish production.

Hence, this view, this vision of directions and trends in process design through a computer integrated process system, directly addresses one of the major cost drivers in the creation of alternate routings and creates two benefits rather than havoc: (1) alternate routings with pure process requirements present the most accurate facility requirements, and (2) alternate routings that are firmed and accurately reflect how a part shall be made can now serve as the basis for absolute shop floor data integrity and configuration management. Further, the cost projection can be accurate for material, indirect labor, direct labor, and capital.

In conclusion, the directions and trends in process design must support the defined needs of tomorrow by allowing flexibility, efficiency, and effectivity in a coexistent manner in the same systems architecture. The architecture can now be achieved with modern object-oriented methodologies, group technology techniques, and data base storage methodologies. Finally, accurate planning and control can be achieved in the budgeting, cost tracking, and material tracking function. In short, the view of computer integrated process is crucial for factory of the future success.

## REFERENCES AND RECOMMENDED READINGS

1. Daniel Appleton, "The State of CIM," *Datamation,* December 16, 1984.
2. John Ettlie, "Managing Design Systems in Manufacturing," NSF Project No. DDM-9007043, 1991.
3. Thomas G. Gunn, "The Mechanization of Design and Manufacturing," *Scientific American,* September 1982.
4. USAF Integrated Computer Aided Manufacturing Report, Project 1105—Factory of the Future, Vought, 1984.
5. A. J. Roth, Jr., "Productivity by Design—FMS Applications That Work," Dearborn, Mich.: FMS Conference, Society of Manufacturing Engineers, 1986.
6. Eli Goldratt, *The Goal* (Croton-on-Hudson, N.Y.: Northriver Press, 1986).
7. A. Greene, "Factory Management Systems," *Production and Inventory Management Review,* May 1988.
8. 1988 Survey, American Production and Inventory Control Society, Falls Church, Va.
9. Dan Shunk and David Peck, "The Creation of a Knowledge Worker Workbench Framework with the Object Oriented Paradigm," Bordeaux, France: IFIP Conference, October 10, 1991.
10. Robert Kaplan, "One Cost System Isn't Enough," *Harvard Business Review,* January–February 1988.
11. S. McCarron, "Open Systems—Evolution or Revolution?" *UNIX Review,* November 1990.

# CHAPTER 10
# DIRECTIONS AND TRENDS

It is clear, from the previous chapters in this text, that process design has fundamentally changed in the past few years. In the past, we focused on speeds and feeds, depths of cut, process planning, variant and generative means by which we could create a process plan. Today, we look at process design as an integral link to the integrated product-process definition environment where cross-disciplinary teams of talented individuals are coming together early in the design cycle to conceive of the product and then carry out the overall product-process plan throughout the entire life cycle of the product with integrated process design as a fundamental link in this overall trend.

To assess where the direction and trends for integrated process design and development are going, we must first look at the issues that are related to integrated process design and then attempt to glean some inferences from that. Early in this text, we talked about the transition from being either efficient or flexible to being efficient and flexible at the same time as we develop a product and a process. This led to the idea that as we discuss strategies for corporations, the notions from Porter and his strategic initiatives of differentiation, low cost, and focus have added four new elements to the strategic thrusts necessary to compete for sustainable competitive advantage in the 21st century. These four are based on the notions of core competencies, from Prahalad; time compression, from Stalk; strategic relationships and strategic alignment due to focus and due to leveraging opportunities, from Skinner; and continuous improvement based on notions all the way back to Deming but now carried forth in works from Drucker.

These four issues are now realigning where integrated process design and development is going. Very briefly, core competencies deal with building world-class leadership in the design and development of a particular class of product and process based on the ability a company possesses. The enterprise competes on the basis of a distinctive set of skills held by the work force. This unique approach as presented in the early 1990s has led many corporations to rethink the way they present strategic plans.

Competencies can be described as having the following characteristics:

1. They must be nurtured and protected because they are the glue that binds the business units together and the engine of the corporation's new business development.
2. They embody the collective learning in the organization and require efforts in coordinating diverse production skills and integrating multiple streams of technologies.
3. They require communication, involvement, and deep commitment to working across organizational boundaries.
4. The organization must be organized around delivering value, which requires being aware of customer's needs and new technological possibilities.

With respect to time, Stalk and others in the late 1980s identified the implementation of time-based management as being the new source of competitive advantage. The corporatewide implementation of this concept will require significant structural changes and new measures of organizational performance. The objective with time-based management is the ability to understand where value is added in the overall product-process life cycle, to focus on the value-added process, to reduce all non-value-added functions to zero, and to strive to an economic order quantity approaching one.

With respect to relationships and the ability to integrate horizontally across the entire enterprise by linking to customers and to suppliers, many have recognized that a horizontal strategy provides for explicit coordination of requirements among business units that makes corporate strategy more than the sum of its individual business units. By being able to develop strategic relationships and fundamentally link one organizational element to another, one can then lever based on the explicit technological advancements and core competencies from one business unit that can go across multiple units within the organization.

Finally, with respect to continuous improvement, Deming showed the way in this area in the early 1970s and has led us into the notion that to be better at what we do, the big bang of having a major splash of improvement often goes astray. But the continuous improvement mode of developing the "as is" understanding and identifying where opportunities for improvement lie on a continuous basis and then dedicating the organization to this activity have led to substantial recognized improvements. Intellectually simple tools are needed here for all involved personnel to gain an understanding of the master plan. Two tools were introduced earlier in this text. The first is a simple tool used by Digital Equipment Corporation and others called "A $\Delta$ T"™ that simply is a process map that captures value-added and non-value-added activities. The second was the IDEF0/td methodology that is a bit more rigorous but can serve as an analysis tool as well as a system definition tool. To support this activity, the model of the business must be created, as we've attempted to do in this text, and then where value is added in the process must be assessed.

To look at these directions and trends, we shall look at the following technological topics as they relate to integrated process design and development and attempt to define briefly the directions and trends that each of these are taking in the future.

## SECTION 10.1 CONCURRENT ENGINEERING

Figure 10–1 depicts the transition in the organizational perspectives of how management and the team must participate in a concurrent engineering environment. As we've discussed previously, a fundamental shift is occurring in the transition to concurrent engineering. Instead of developing the product and then the process in a serial fashion, the teams of talented individuals from multiple disciplines come together around a "roundtable" early in the design cycle to develop this product-process definition, such that it is profitable, able to be fabricated, able to be assembled, producible, and supportable in a cost-effective manner that ultimately meets the customer's requirements. In the evolving global companies, this roundtable may be a virtual table that is spread throughout the world through the electronic linking of corporate centers of excellence. This electronic linking through a corporatewide electronic data

## FIGURE 10-1
## The Transition of Responsibility in Concurrent Engineering

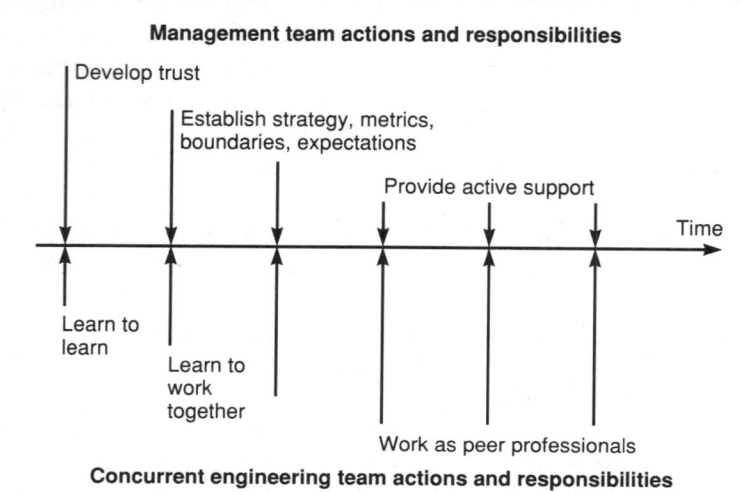

**Management team actions and responsibilities**

- Develop trust
- Establish strategy, metrics, boundaries, expectations
- Provide active support

Time →

- Learn to learn
- Learn to work together
- Work as peer professionals

**Concurrent engineering team actions and responsibilities**

interchange (EDI) and/or a technical data interchange (TDI) network provides the advanced technology mechanism to allow this to succeed.

The fundamental transitions we see in concurrent engineering deal primarily with technological advances and also cultural advances necessary to support these notions. From the cultural standpoint, Figure 10-1 depicts the transitioning that occurs over the time line of the concurrent engineering framework. Early in the concurrent engineering role, management must develop the overall strategies and boundaries of the teams and instill on the individuals and teams the notion that trust is present and that the overall metrics for the team are truly the metrics by which the benefits of the overall organization must abide. After this transition occurs, the management team continues by developing the metrics and the boundaries for the teams and the clear recognition that they are developing support for this overall teamwork environment. This must be an overt action on the management team's part to demonstrate how this teamwork environment will be supported.

As the time line progresses, the team takes over in its responsibilities to define ways of working together; this team may be a global team with individual members located throughout the world. As they develop their own style and clearly define the mission of the overall team agreement, the team must do the following:

1. Understand what to do with the individuals who have not had that much experience in working in a team environment.
2. Understand what to do with the individuals who wish not to work in a team environment.
3. Address the overall objectives the management has placed on the team.
4. Get on with the product-process development cycle.

The culture must shift to support this transitioning to the team. Many companies are recognizing that to be successful, authority and responsibility typically handled at one level within the organization must shift. In the past, authority was often held at the upper levels within the organization and responsibility was distributed to lower levels. However, in today's environment, just the opposite is occurring. Authority is now distributed throughout the teams, through charters, strategies, metrics, boundaries, and overall charges and mission statements to individual team members. The management team must maintain the responsibility to bring the team together and to provide the resources to make this overall notion happen. This fundamental shift is causing major cultural changes, as Savage and others have pointed out in their late 1990 texts about cultural change in the integrated environment.

To support this concurrent engineering environment, four fundamental shifts are also occurring in the areas of multimedia presentation, the ability to operate in a global environment, and the ability to do inferencing from the data bases that exist today. First, in a multimedia presentation, the planners are not just dealing with blueprints in today's global product-process development cycle. The planners are attempting to develop text, glossaries, audiovisual presentations, voice, video, and full data representations of products and processes for the ultimate customer understanding and acceptance of their efforts. The multimedia presentations of the 21st century will be fundamental to this success. However, we are just beginning to understand from a technological standpoint how to integrate voice, video, and data and many other data base representations in one methodology, called multimedia presentations.

One notion in this area is called "hypertext." In hypertext, the technological advancements of voice, video, data, data management, and other issues are being clearly represented in an attempt to present the very best information to the product-process development team as well as to the customer and the supplier in order to make the product-process cycle flow smoothly and be correct.

The second technological advancement deemed necessary is the ability to handle concurrent engineering product-process development in a global environment. This will be called "passing the chalk" because the fundamental notion here is the ability to have a single representation of the product-process carried out in multiple locations throughout the world and then to "pass the chalk" in an integrated fashion such that each team is working on a single representation on the product and process rather than individual representations. The ability to pass the chalk has been represented through technological advancements in 1991 in various laboratories throughout the United States. It requires a fully networked capability throughout each of these design sites that allows the management of time based on a 24-hour clock because many of these sites will be around the world. It also allows the system to communicate on a real-time basis when communications are required. The notion of passing the chalk is an advancement that we believe is fundamental to the overall success of the concurrent engineering roundtable whether it be at a single site or at multiple sites throughout the world.

The third notion of directions and trends of technological support as they relate to concurrent engineering deals with inferencing from data base content. We must understand how to gather requirements from the customer, how to preserve the data of this activity, and how to present these data to the product-process development teams. How does one provide access to all the information that has been developed on previous development teams, and how does one cross-reference and develop interlinking methodologies such that the data that existed in the past could be utilized to help the data development necessary in the future? This "mining of a data base" is the ability to extract from alien data bases the information necessary to glean inferences in a very, very rapid fashion and develop abilities to infer what one ought to do in this particular situation to support the notions developed in the past.

The technological issues here are that we must understand the distribution of data. How does one distribute data, how does one maintain configuration control, and how does one then handle the overall abilities to provide access to the data? Also involved is the archiving for heterogeneous environments, where we recognize that many companies, probably all companies, have multiple computing environments and multiple data structures. Planners are recognizing that to compete in the 21st century they must access multiple data bases to be successful for this overall product-process development cycle. The archiving and the ability to peruse in a heterogeneous environment are nontrivial notions that we recognize as being fundamental to the success of the overall product-process development cycle. To a limited extent, this technology is here and is being technologically expanded in a rapid fashion to true commercial viability.

The fourth issue in the shift toward a concurrent engineering environment relates to the advancements made to support the user as he or she interacts with the systems. The user interface must be tuned to the capabilities of the work force. Research shows that the user interface must be based on the familiarity of the user to the systems.[1] Current computer technology has caused many workers to become computer users. However, many do not feel comfortable with the systems of today. The development of the user interface that mediates between users and applications has risen in importance because of the focus on the people that truly add value to the product. Recent research in the cognitive issues of the workplace include how people come to understand the goals they are trying to achieve, how they adapt to changing conditions, how decisions are made, and how operators achieve integrated and coordinated actions. Other issues include the problems of multiple users, user learning, user preference, and proper help functions. Research into the framework of an advanced tool that addresses the aforementioned issues is ongoing at many institutions. The requirements are:

1. The tool will identify the user in terms of user knowledge level and recent skill set.
2. The tool will provide a number of dialogue types and offer

the user a proper dialogue type according to the user preference and knowledge level.
3. The tool will be an intelligent interface.
4. The tool will provide enough help to the user.
5. The tool will not only provide consistent interactions between knowledge workers and applications but also process the assigned tasks concurrently.

Such elements as touch screen, voice activation, full-motion video, windows, multilanguages, and headset operation rather than keyboard operation are all being developed as mechanisms for this cognitive interface.

Finally, the impact of concurrent engineering emphasizes the design through the build integration cycle of the overall product and process. The impacts of these and the needs have led us to new abilities to access this information, possibly through new user interfaces that recognize the needs of skilled process or product development users who are novice information access users. Novel user interfaces are being developed that allow inferencing and mining of this rich data structure in a very efficient manner.

New access methodologies must be developed that also attempt to develop layers of information. Systems will need to provide the ability to access bits and pieces at a higher, macro level so that the teams can recognize whether this data stream has value. This layering notion is new and is a direction and trend that we must understand in terms of how to peel back the "onion skin" in the overall recognition of where the data are that are necessary to develop the overall product-process cycle.

## SECTION 10.2  TOTAL QUALITY MANAGEMENT DIRECTIONS AND TRENDS

We are developing this integrated process design and development notion in a total quality management environment; hence, the notions of modeling the overall product-process development cycle were introduced early in this text. Total quality management in the mid-1980s was simply the capturing of X-bar and R charts and developing statistical trends over the life cycle of a product. This was called "statistical process control."

By definition, new-product development will not be the same as previous products. However, in Japan, the notion of product-process is not just focusing on the end result; rather it is focusing as much on the *process* by which the new product was developed. The Asian Rim nations have shown the rest of the world how the implementation of total quality management and the philosophy of continuous improvement can be exploited and utilized effectively. TQM reveals that the process for product-process development is as important as the end result. This is fundamentally shifting the overall metrics of many corporations in the United States to be able to recognize and reward teams of people who develop an improved process even though their product may not be a world beater in the overall cycle. Philosophically, this is significantly different from how we acted in the past.

Total quality management requires a simple process methodology with a clear vision and understood principles, practices, and techniques. Tools are needed that can aid the TQM in the recognition and development of a map by which existing processes are performed and include sufficient feedback such that improvements can constantly be made. We have introduced in this text one or two different methodologies to begin the process of recognizing how these overall product-process development cycles can be modeled.

The second important advancement in TQM for the integrated enterprise of the 21st century is the ability to glean inferences on product and process quality from small lot sizes. By introducing the simultaneous needs for efficiency and flexibility, lot sizes are becoming smaller. But many of the statistical tools need larger sample sizes (e.g., 40 to 100 data points) to gain some confidence in the results. Research into the application of group technology techniques for small lot SPC is now evolving. By taking advantage of the process similarities across parts in a family, SPC inferences can be generated even if the individual sample sizes are too small.

## SECTION 10.3   MATERIAL SUPPLIER INTEGRATION

The notion that we are developing strategic alignments with organizations includes strategic alignments with our supplier base. Our material suppliers today are becoming strategic partners in our overall corporate horizontal integration.

To achieve strategic alliances with our suppliers, they must be brought in early in making decisions, and we must be able to interact with their corporate knowledge and embellish that in terms of development of the overall product and process within our corporation. The material supplier integration is fundamentally a linking process that is secure, is controlled, is managed, and allows the supplier base to maintain its security and the company to maintain its security and yet tap into one another's information banks to truly understand where the technology is going and to be able to leapfrog to the next generation of product.

Often, in today's environment, the material supplier's breakthrough can lead to a new product and process that the individual companies were unable to define on their own. For example, in the computing industry today, the chip manufacturer has built functionality into the new microcomputer chipsets based on customer requests. Motorola, Intel, and others that have let new generations of product provide the computer makers and other electronic equipment makers with exciting new opportunities. The Motorola 68040 chips and the Intel 486 chips have had dramatic impacts on the overall opportunities of many companies throughout the world.

For this new-product development at the supplier level to be successful, the supplier must be integrated with the overall product and process at the user level. This is a nontrivial exercise because most of the time these product and process development cycles are going on simultaneously even at the supplier level. Hence, strategic alignments within the material supplier base are fundamental in the networking, communication, and accessing of proper data and the development of a neutral file, if you will, because communication between supplier and customer is fundamental to the overall success of the integrated product-process development cycle.

One of the most unique approaches to supplier integration has come from a supplier. A job shop in Phoenix has established a world-class network of casting houses, forging houses, and raw material suppliers. By doing so, the job shop is now in a position to offer "one-stop, world-class shopping" to the aerospace prime contractors. By assimilating the supplier base core competencies under one banner, the job shop can minimize the prime contractor's risk and add significant value to the product while maintaining minimal overhead.

A second example is an American manufacturing facility that ties the statistical process control (SPC) measurements, by lot, to a supplier's product or an internal lot, so that when the just-in-time lot hits the next critical operation on the factory floor, the tooling on the floor has been automatically calibrated for that lot.

## SECTION 10.4   EQUIPMENT SUPPLIER INTEGRATION

The directions and trends that are seen in equipment supplier integration begin with the recognition that a technological food chain exists throughout all of manufacturing. For the enterprise to produce final products for customers, it must have intermediate products, or subassemblies, that typically come from material suppliers, and it must also have basic equipment advancements that come from equipment suppliers in order to develop the overall vertical and horizontal integration of our enterprise.

Figure 4-8, shown earlier, depicts what is colloquially called the technological food chain. This was developed by a colleague at Motorola and depicted to show how the overall product-process development cycle occurs from a technology standpoint. At the highest levels within this inverted pyramid are the final products. The intermediate level depicts the subassemblies and commodities necessary to build that final product. These might be the computer chips to build the personal computer, the composite materials necessary to create the next generation aircraft, and so on. At the base of the pyramid is a fundamental understanding of how the equipment needs and the basic material needs must be met to succeed in this product-process development cycle. Often, the next generation equipment defines the next generation subassemblies, which define the next generation final assemblies that you and I use.

Equipment suppliers have affected the success of the product-process integration. Digital Equipment Corporation had tremendous success in the integration of the conveyor and automation equipment in the design and production of the RA90 family of automated digital storage devices. Because of this and other integration successes on this line, the RA90 won the Society of Manufacturing Engineer's LEAD Award in 1989.

The equipment supplier base in many countries has been eroding. It is not a glamorous industry. Machine tools for metal cutting and metal removal, lithographic equipment as a fundamental building block for semiconductor fabrication, precision machinery for gear development and ball bearings, have often been looked on as mediocre, less glamorous industries with small profit margins, high risk, and small volumes. Many politicians and businesspeople have come to believe it is not necessary to maintain a strong equipment supplier base to achieve fundamental advancements in the overall product-process development cycle.

However, it is now clear that for vertical and horizontal integration to be successful in the product-process domain, the abilities of equipment suppliers to join partnerships is fundamental to the success of the overall enterprise directions.

## SECTION 10.5  PROCESS PLANNING DIRECTIONS AND TRENDS

The directions and trends of process planning are directly linked to the following advancements:

1. The migration to knowledgeable product definition systems that contain robust understanding of the product rather than "stick figure graphics" of the 1980s.
2. The advancements in the virtual data base that makes the data base appear as a single structure concept for the entire enterprise that will allow access to capacity and capabilities throughout the world.
3. The formal linkages through manufacturing standards of design data structures with planning and bill-of-material data structures.
4. The true formalization of the planning process with legitimate process models that are consistently used and produce a good process plan every time.
5. The advancement of the artificial intelligence and neural network support technologies in order to capture the logic and effectively use it.

6. The interactive simulation capability that extracts current status from the factory floor and allows "what if" analysis with good data.

Given all of these advancements, expected in the next 10 years, the "art" of process planning could then become a "science" and an integral link in the product-process design. Through this integration, the linkages to flexible equipment can be achieved that will allow the downloading of one job at a time without the setup and teardown penalties of today. A critical element of this integration shall be the computer-aided tolerancing and dimensioning capability. Tolerance stacking and dimensioning is not a mature technology. By having this tool in the planner's arsenal, the 10-step process outlined in Chapter 5 can be achieved in an advanced CAPP system.

Finally, the transition to the team concept will allow the focused, modular cells of the factory floor to operate with cross-functional teams that need a "knowledge worker workbench" set of tools by which to run their "microstrategic business unit."[2]

## SECTION 10.6   PROCESS CONTROL DIRECTIONS AND TRENDS

The trends in process control focus on the ability to characterize the process and the product in such a goal-oriented fashion that self-direction is possible. By self-direction is meant, "systems that have the ability to reason, i.e., to make process-path decisions during the process."[3] Dr. Steven LeClair is one of the foremost developers of self-directed process control. In his recent article, he introduces the notion that intelligent manufacturing involves intelligent, product-process-oriented systems that can facilitate the on-line, real-time generation of an improving product-process cycle. The state of the art is to create systems that can self-direct and self-learn. These result in improvements in quality and reliability plus improved cycle time—all leading to improved productivity.

The technological advancements are in the areas of "goal-driven" versus "state-driven" process. LeClair differentiates between "goal space modeling," which looks at the process as a series of events, or goals, regarding the product during the process, and "state space modeling," a function of the process time irrespective of the product. The difference is the ability to sense how the product is proceeding through the various processes and to *adapt* for the optimum process. The term for this is *qualitative process automation* (QPA) given to this research focus by the U.S. Air Force's Materials Laboratory at Wright-Patterson Air Force Base. Results from the lab's research have been dramatic. It reports that the fluctuations of the source material temperature in a nonmetallic processing system was reduced by 50 percent. Clearly this direction and trend requires research into the need for integrated systems, but the results are significant.

## SECTION 10.7 ASSET MANAGEMENT AND CAPACITY PLANNING IN THE INTEGRATED PROCESS DESIGN

As we discussed earlier, transitions in asset management and capacity planning have fundamentally changed the way inventory is viewed and factories are scheduled. These new methods are based on new management strategies, such as those of time-based management espoused by Stalk and others. To achieve customer responsiveness with an economic order quantity of one, traditional systems of large production runs have migrated to new philosophies where rapid response to changing customer demand becomes the focal point for proper asset management. This, in certain cases, demonstrates that capacity planning must often be planning flexible facilities as compared to planning fully utilized facilities, an example being the Allen Bradley facility in Milwaukee, Wisconsin. Allen Bradley won the Society of Manufacturing Engineers LEAD Award for excellence in the application of CIM in 1985. Due to world competition, it was forced to make a major transition in its philosophy and capacity of producing starter motors to compete.

Originally, the plan was to produce approximately 100 different types of starter motors in the facility in Milwaukee. However, Tracy O'Rourke, then chief executive and president of Allen Bradley, and his planning staff found that to be responsive the company had to produce a fully integrated, fully automated facility that could produce over 800 different varieties with a response time of 24 hours. In this fully integrated, fully automated environment, the product-process development cycle became integrally linked to the overall capacity planning and asset management philosophies. And, in this case, capacity planning turned out to be one of designing a flexible integrated system as compared to a fully utilized system. Results have been phenomenal. Tracy O'Rourke and others from Allen Bradley report they have a "financial superstar," with a capacity that is utilized at only 40 percent while producing an order in 24 hours with a lot size of one of any of 800 different varieties of a product.

New philosophies of facilities layout have led to the recognition that for proper asset management and capacity planning in today's environment, we must recognize that the traditional low-cost, high-volume strategies have changed. Managers in the past had to do whatever it took to drive down costs. This is no different from today. But the approach in the past differs from today's. For example, in the past, moving production to low labor cost countries, contracting out to lowest cost vendors, gaining economies of scale by building new facilities of larger scope, and focusing operations on the most economic subset of the activities were used. But today, managers are changing their philosophies to be able to define and fully understand how to compete in a time-based environment.

## SECTION 10.8  RAPID PROTOTYPING

The need to rapidly capture concepts in a form that the user can see and feel is gaining importance. This need has stimulated a whole new industry that deals in rapidly prototyping concepts. Most of these prototypes fall into the category of "proof of concept" and "proof of product" as developed in Chapter 5. The objective of this technological advancement is to:

- Provide user access to concepts early in the design phase.
- Facilitate the integration of product and process.
- Stimulate new thought on how to improve the product.
- Disseminate the ideas with physical representations.

The effect of rapid prototyping in process design and development occurs when the latter two prototyping stages are incorporated into the overall plan, these being "proof of producibility" and "proof of production." Nonetheless, any prototyping methodology that can speed the product realization process is being reviewed at all major corporations today. In a growing number of industries, ranging from microelectronics through consumer electronics to automobiles, product life cycles are shrinking. These are now measured in months rather than years. Rapid product realization determines the enterprise's ability to meet market demand and respond to changes in that demand precipitated by customer changes or competitor's initiatives. Rapid prototyping, when prototyping is warranted, is a major new direction for product-process integration in the 21st century.

## SECTION 10.9  FLEXIBLE, FOCUSED FACTORIES FOR INTEGRATED PRODUCT-PROCESS DESIGN

Dr. Wickham Skinner, in his classic article, demonstrated the importance of manufacturing and recommended that flexible focused factories be the building block on which the modern manufacturing scenario be built.[4]

To do this, we are beginning to recognize, as we notice the directions and trends to compete in the 21st century, that we are constantly striving for new and novel ways to make the customer more successful and to be responsive to changing customer demands. The layout of processes in a facility differ from the past in that the traditional process centers organized around the type of machinery has changed to product centers in flexible environments using group technology as the foundation. Also, production scheduling in the past was typically performed in a centralized, sophisticated material resource planning and shop floor control system; whereas in today's environment, with rapid response being the byword and decisions being pushed to the lowest level within the manufacturing organization, time-based manufacturing relies on localized scheduling. Working in concert with product center layouts, these modules of manufacturing make the total production process run more smoothly using demand-pull asset management abilities.

This means the product-process development cycle must now

be planning products to be produced in a virtual environment anywhere in the world, and philosophies such as "globalization with localization" are now fully capable of being realized. By globalization with localization we mean that many organizations in the world today are global in scope, possessing abilities to present products to market throughout the world and to define and have the strength and the leverage for worldwide access to talent, equipment, materials, and so on. However, this also means that through the localization notion, it looks as though the factory is right next to the customer base and through flexible, focused factories can respond very rapidly to changing customer demands. And it appears as though a totally custom product is developed at the customer site and only for this one customer.

This migration to globalization with localization is one of many fundamental changes occurring as we have become totally customer oriented in the late 1980s and are migrating into a full-fledged customer-oriented society in the manufacturing environment in the 1990s.

## REFERENCES AND RECOMMENDED READINGS

1. Seouk Joo Lee and Dan L. Shunk, ongoing research.
2. Dan L. Shunk and David Peck, "The Creation of a Knowledge Worker Workbench Framework with the Object Oriented Paradigm," Amsterdam: CAPE '91 Proceedings, IFIP, Bordeaux, North Holland Press, 1991.
3. Steven LeClair, "Self-Directed Process Control," *Concurrent Engineering,* July–August 1991.
4. Wickham Skinner, "Manufacturing—Missing Link in Corporate Strategy," *Harvard Business Review,* May–June 1969.

# CHAPTER 11
# AN INTEGRATED PRODUCT-PROCESS CASE EXAMPLE

Reflecting on the initial concepts of the integrated product-process design text, we again look at Drucker's notions that the new measurement unit for the world-class manufacturers of the 21st century must be *time*. Although we haven't built it yet, we can already begin to specify what that postmodern factory of 1991 is going to be. But global competition is real, and in many cases, as Mr. John Young and others reported in their Presidential Commission Report in 1985,[1] the wealth-generating capability of nations is shifting, and that wealth generation is fundamental to the success and prosperity of nations. There are only three ways to generate wealth, as Young points out in his report. You either:

- Grow it.
- Extract it.
- Make it.

The focus of this text has been on the making of products to generate wealth for nations by providing a valued service to the customer.

In looking at where wealth is being generated in integrated process design success stories around the world, some examples are presented of how the Japanese are developing integrated process design and development as a subset of their overall product-process development cycle.

Professor Joseph Kimura at the University of Tokyo states that the survival strategy for advanced industries within Japan is based on three relatively simple yet highly productive notions. First, they must produce high-quality, value-added products. Second, they must invent new products in such a timely fashion that members of the buying public may not even know they need the product before they see it. Third, they must be able to produce extremely small production lot sizes, as small as a lot size of one when needed. These three simple notions have been reflected in this text.

High-quality, value-added products must develop in an integrated product-process development cycle where concurrent engineering, total quality management, supplier integration both from a material standpoint and equipment standpoint, proper asset management and capacity planning, the ability to rapidly develop a prototype and rapidly develop time-to-market compression, and finally, the notion of flexible and focused factories have just been exploited in Chapter 11. All of these are fundamental to this survival strategy, whether one is in Japan, the United States, or anywhere else in the world.

The Japanese have passed the United States in total capital expenditures in 1989 and again in 1990. Their corporations spend 30 to 60 percent more on research and development than do those in the United States and around the world, and Japan now controls the world's top 12 largest banks and 17 of the largest 18 banks, according to *Business Week*.[2] The control of capital, the greater emphasis on research and development, and the greater capital expenditures have led Japan to a very dominant position in many of the world marketplaces as reflected in the report *Made in America*.[3] In this report, Japan now controls the propensity of the marketplace in areas such as consumer electronics, automobiles, and more.

The Japanese are also developing some novel approaches to integrated product-process design. They are creating virtual manufacturing factories to minimize the time to market. As reported in the Nogoya Conference of the U.S.-Japanese Manufacturing Technology Exchange in June 1990, Hitachi can simulate not only the product but also the processes required, and through this simulation it has significantly reduced the time it takes to develop and get a product to market.[4] Also, the Japanese have demonstrated that if we look at all the product-process development cycles, huge amounts of data are being generated. To analyze and understand how these pieces of data can be used, sensor fusion and neural networks, advanced technologies using artificial and neural network types of approaches to life, are being developed.

Finally, the Japanese can rapidly develop a product and get it to market. This is important, as Figure 11–1 shows. Based on the *Electronics Business Week* report of May 1989, the figure shows that in the dynamic random access memory market, the advantages of being first are quite obvious. As we have migrated from various

**FIGURE 11-1**
**Results of Time-to-Market Leadership**

| Product Type | First Company | Year Introduced | First Year in Production | Market Leaders in First Year | Percent of Market |
|---|---|---|---|---|---|
| 16K | MOSTEK | 1976 | 1978 | MOSTEK (*NEC* | 25 *20%)* |
| 64K | Hitachi | 1979 | 1982 | Hitachi (*NEC* | 19 *15%)* |
| 256K | NEC | 1982 | 1984 | NEC (*Hitachi* | 27 *24%)* |
| 1 megabit | Toshiba | 1985 | 1987 | Toshiba (*Mitsubishi* | 47 *16%)* |

product types of 16K, 64K, 256K, and 1 megabit dynamic random access memories (DRAM), the first company to develop that product has normally taken approximately two years from when it was introduced to the first year it was in production. Yet, that first year in production, the first company to introduce the product has gained significant market share. Remarkably, Toshiba in the 1 megabit DRAM area in 1987, captured 47 percent market share in a premium-priced product based on its ability to rapidly exploit time to market.

This can be coupled with the results George Stalk references in his time-based competition series when he compares a Japanese company's and an American company's ability to develop a product and get that product to market in a relatively short time. As we see in Figure 11-2, the Western company for this particular mechanical transmission took 38 months to get that product to market. Whereas, the Japanese company with a very similar mechanical transmission took only 19 months.

But as we look at this, we notice two very startling effects. First, wherever a mechanical type of thought pattern is required, such as layout and detailed design and the interpretation of the manufacturing prototype and development, and then prototype testing, the Japanese are very efficient at what they do. They are very, very fast. Second, in thought patterns where consensus is required, such as design reviews, the Japanese take significantly longer than the Western company in performing that function. Here they are slow.

The analysis of this leads one to conclude that "hurry, hurry, hurry" is not necessarily the best notion to employ when we talk about time-to-market compression. There will be certain times where integration of systems will lead the company to a very, very efficient activity, such as layout. However, where thought patterns are required, *slowing the process to speed up the overall product delivery* is an intellectually inconsistent, yet a fundamentally sound approach to development of time-to-market compression. As Socrates said, we must "hurry slowly," and in the design review process, where individuals of various functional elements within the organization get together to agree that the product is ultimately ready for production, that is the case.

**FIGURE 11–2**
**Time-to-Market Comparisons**

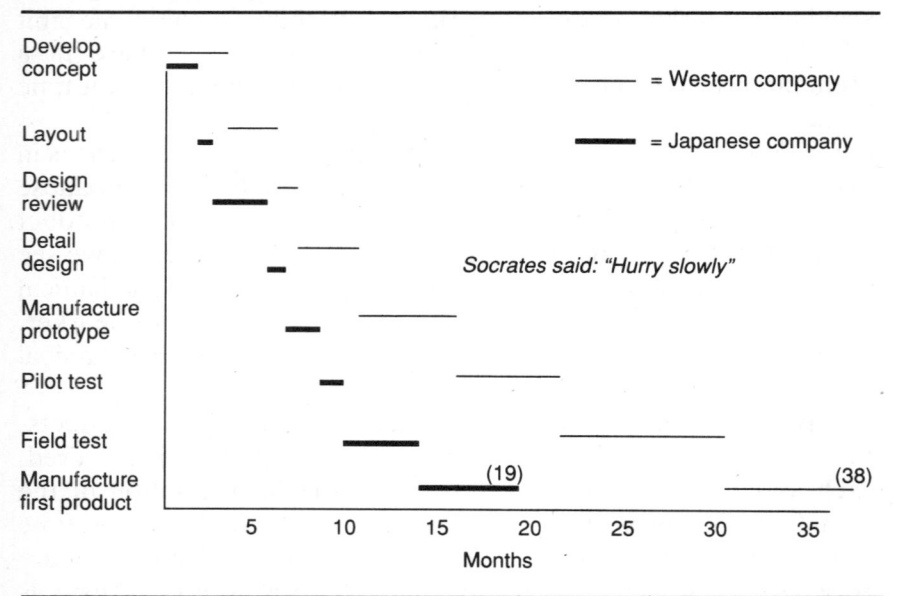

*Socrates said: "Hurry slowly"*

——— = Western company
▬▬▬ = Japanese company

|  | |
|---|---|
| Develop concept | |
| Layout | |
| Design review | |
| Detail design | |
| Manufacture prototype | |
| Pilot test | |
| Field test | |
| Manufacture first product | (19) ... (38) |

Months: 5, 10, 15, 20, 25, 30, 35

## SECTION 11.1 AN ASSESSMENT OF INTEGRATION IN JAPAN

Each of the case studies in this book begin with the results of a recent survey related to the case. In this chapter, we look at the novel ways the Japanese are integrating the process with the product. The reference here is a survey published in 1989 by the American Society of Mechanical Engineers.[5] It surveyed 284 individuals from six industry categories in Japan:

- Precision machinery.
- Transportation.
- Metal products.
- Electrical and electronics.
- Industrial machinery.
- Other manufacturing.

Respondents held all levels of management and came from most professions involved with manufacturing, with the majority coming from planning, management, and manufacturing. Key to the survey are the results shown in Table 11–1. The survey focused on the impact of short-term profit on the investments in FA (factory automation). In Table 11–1a, note that *only* 8.4 percent of the respondents agreed that the investment must concern itself with short-term profit, while in Table 11–1b note that 58.8 percent of justifications are for longer than three years. As is noted in the Nippondenzo case that follows, the Japanese tend to invest in systems for purposes other than short-term profits. And their results are so successful in many cases that all the world is taking note.

## SECTION 11.2 THE NIPPODENZO CASE STUDY

One particular example, that of Nippondenzo, the plant works in Anjo, Japan, shows how a product-process development team can exploit the notions presented in this text. In Anjo, Nippondenzo was gracious enough to participate in a National Research Council study of the United States dealing with time-to-market compression

**TABLE 11-1**
**Survey Results of Japanese Investments**

| Response | Total | Percentage |
|---|---|---|
| *a. The Impact of Concern for Short-Term Profit on Investments* | | |
| Strongly agree | 2 | 0.7 |
| Agree | 22 | 7.7 |
| No opinion | 65 | 22.9 |
| Disagree | 169 | 59.5 |
| Strongly disagree | 14 | 4.9 |
| No answer | 12 | 4.2 |
| Total | 284 | 100.0 |
| *b. The Expected Payback Period of a Typical Investment* | | |
| 2 years | 12 | 6.3 |
| 3 years | 67 | 34.9 |
| 4 years | 19 | 9.9 |
| 5 years | 64 | 33.3 |
| 6–10 years | 26 | 13.5 |
| Other | 4 | 2.1 |
| Total | 192 | 100.0 |

and rapid prototyping in 1989. In 1989, Nippondenzo had a 28,000-square-meter prototyping facility. It also had a 50,000-square-meter production facility. This plant dealt with automotive components where the automotive components were typically in an *evolutionary* state. This is not a revolutionary type of product development cycle, but it supports the evolutionary product development cycle found quite often in many of the world's major markets.

In the prototyping facility in Nippondenzo, *fully characterized production equipment* was being utilized to develop prototypes. Compared to American and other world standards for effective utilization of capital equipment, this is strategically inconsistent with the way a company should develop a product. Normally, prototyping facilities use breadboard types of pieces of equipment where the product is developed without truly testing the overall process.

Nippondenzo has shown, as depicted in Figure 11-3, that a cost-effective way of investing capital resources is to effectively develop product and process simultaneously during the prototyping

**FIGURE 11-3**
**The Nippondenzo Facilities in Anjo, Japan**

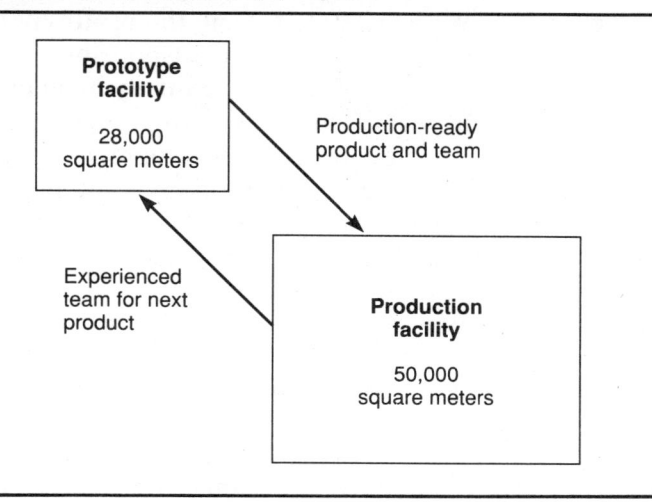

```
        ┌──────────────┐
        │  Prototype   │
        │  facility    │      Production-ready
        │              │──┐   product and team
        │   28,000     │  │
        │square meters │  │
        └──────────────┘  ▼
           ▲        ┌──────────────────┐
  Experienced       │                  │
  team for next     │    Production    │
  product           │    facility      │
                    │                  │
                    │     50,000       │
                    │ square meters    │
                    └──────────────────┘
```

phase with a team that is measured not by the design of the product but by the overall market share and the overall time-to-market compression the product achieves.

As Nippondenzo develops a product in the prototyping facility, the tooling, the fixturing, the strategies, the processes, everything is developed simultaneously with the overall product. As the team and the processes migrate from prototyping facility to production facility, the time between prototype and production is minuscule. It could be measured almost as though it's one weekend—you leave the prototyping facility on Friday, and on Monday morning, the overall concurrent engineering cross-functional team migrates to the production facility to be responsible for overall production capacity. The team is made up of all aspects of workers that represent their functional elements in the overall product-process development cycle. The design engineer is an element of the team, and also the manufacturing engineer, the tooling engineer, the materials engineer, the supplier representative, the customer representative, the systems representative, and so on. They all make up this cross-functional virtual task team responsible for the overall product-process concurrent engineering environment.

As this product matures to the point where it is ready to go into production, the cross-functional team migrates with the product over a weekend, as we have stated, from the prototype facility to the production facility. The team has the responsibility to ramp up production and capture market share in a timely manner.

The fascinating aspect of this, as shown in Figure 11–4, is that the notion of *strategic* new-product development is the critical element of this overall strategy. Here, product and process development teams are launched simultaneously in what Nippondenzo called the notion of "Jikigata-Ken," representing strategic new-product development council strategies.

The same cross-functional, virtual task team that launched the product design is also responsible for launching the overall process design in a simultaneous fashion and taking the product to market in a very cost-effective manner. Once the team has been successful, with product ramp-up, assembly operators trained, and the product is doing well in the marketplace, the team is rewarded by having the experienced team return to the prototyping facility for the next product-process development cycle. This circular flow of talent based on an inconsistent investment strategy of locations of capital equipment is a fundamental difference between the way the Japanese have developed their products and Western societies have developed theirs. And the representation of market share and time-

**FIGURE 11–4**
**The "Jikigata Ken" Scenario**

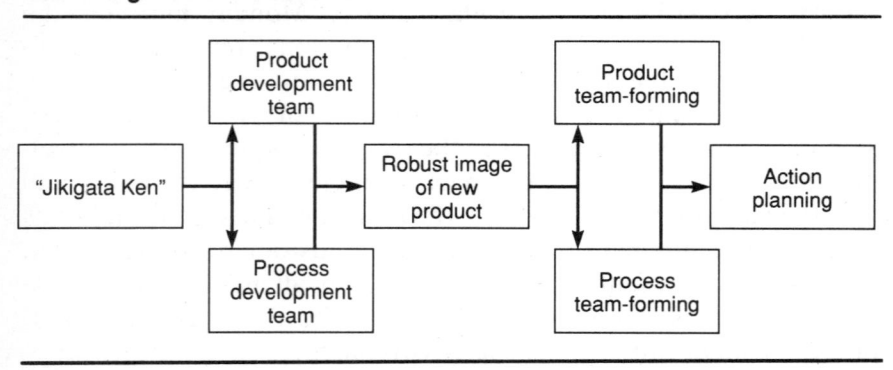

to-market compression that the Japanese have demonstrated bears significant merit in this overall integrated product-process case example.

## SECTION 11.3 NEW METRICS FOR THIS INTEGRATED PRODUCT-PROCESS EXAMPLE

As we've seen, Nippondenzo and others are not using the metrics of today to attack world-class product-process development. New metrics must be applied. The universal metrics discussed in Chapter 7 of "new, good, fast, and cheap" have migrated to different dimensions as we address 21st century manufacturing.

"New" product development is a fundamental element in the metrics of today. "Good" has migrated simply from meeting the customer requirements to six sigma levels of quality. "Fast" means the product must get to the marketplace and may be the dominant factor from a time domain standpoint as Drucker points out. "Cheap" certainly represents the utilization of assets and the proper inventory levels, and so on, as they relate to cost of product.

However, as one can begin to assess how one puts new parameters on these metrics, market share dominance and time-to-market compression are two fundamental issues being addressed in 21st century manufacturing thought. The issue starts with the new notion that says the team must have boundaries and strategies that address not only the new-product design but also the total product-process life cycle development. The overall strategy of how a product is developed must be rethought in due manner.

## SECTION 11.4 THE ROLE OF THE PROCESS DESIGN FUNCTION

The sequential world of the 1980s is migrating to the parallel planning world of the 1990s where Jikigata-Ken in all its glory will be a way of producing product in the 21st century. Jikigata-Ken is not just a process. It is the way by which the entire design function and the teamwork approach are being dictated to achieve world-class standing. Jikigata-Ken states that product and process are conducted in a simultaneous or concurrent fashion and they are conducted by teams of folks dedicated not simply to their functional element but to the overall success of the product and process.

To do this, the role of the process design and development function is changing. It is not a reactive organization. It is a proactive organization that has the ability in conceptual, preliminary, and detail design phases to provide good input into the overall product-process development cycle. It is not simply one that catches a design as it comes tumbling over the wall, but it participates in the overall teamwork effort. The function is being rekindled and totally rethought and Nippondenzo is an excellent example of that.

## REFERENCES AND RECOMMENDED READINGS

1. John Young, "Global Competition—The New Reality," Presidential Commission Report, Washington, D.C., 1985.
2. "The Global 1000—The Leaders," *Business Week,* July 15, 1991.
3. Michael L. Dertouzos, Richard Lester, and Robert Solow, *Made in America* (Cambridge, Mass.: MIT Press, 1989).
4. U.S.–Japan, Manufacturing Technology Exchange Conference Proceedings, Washington, D.C.; Academy Press, 1990.
5. Philip Huang and Michiharu Sakurai, "An Assessment of Factory Automation in Japan: A General Mail Survey," *Manufacturing Review,* September 1989.

# CHAPTER 12
# PRIMARY INTERFUNCTIONAL CONNECTIONS FOR PROCESS DESIGN AND DEVELOPMENT

## SECTION 12.1  LINKAGES FOR PROCESS DESIGN AND DEVELOPMENT

In the format of the CIRM texts being developed for the American Production and Inventory Control Society (APICS), the final chapter in each text is dedicated to primary and secondary interfunctional connections. However, as we've developed the notion of integrated process design and development in this text and have determined that some variation of the concept of concurrent engineering is the direction many major manufacturing organizations in the world are heading, the notion of secondary interfunctional connections dwindles in scope to the fundamental issue that *everyone* will be integrated with *everyone* around the global, virtual "roundtable" to develop product and process simultaneously to meet the customer requirements.

To do this, we must understand the implications for world-class manufacturing. As Dr. Steven Wheelwright of the Harvard Business School related,

> To compete in world class and be a world-class manufacturer in the 21st century, every company must *first* develop a superior product and service in a more rapid manner than they are today. *Second,* they must introduce new manufacturing technologies and approaches more rapidly than they are today. And *third,* they must select and train more capable workers and managers than they are today.[1]

In each of these, the fundamental issue is that people must work together in a much more timely, integrated manner to meet the needs of the changing customer base.

This requirement is fostering change in every element of the culture. The corporations are changing. They are redefining and reinventing the way they do business. The education system is changing to support that. And finally, the government role is changing in an ever increasing recognition that to generate wealth one must manufacture products.

Many companies are finding that to do this they must bring together diverse elements within the organization into a singular entity called the "product-process development team." For example, one major aerospace corporation within the United States has found that for it to be successful, design and drafting must be integrated with planning and scheduling, which must be integrated with fabrication and test to create the integrated enterprise for the 21st century organization. It quotes, "The change in the next 3 years will equal the change that happened in the organization in the last 15."

We are finding that stand-alone equipment is migrating to integrated equipment and systems and that single-discipline employees are migrating to have multiple disciplines and multiple exposures to the various disciplines that exist within the enterprise. This is forcing companies to rethink the overall job skill and job rotation concepts of the past. To successfully create multidisciplined, cross-functional team members, an investment in the human element is necessary. We must truly demonstrate that people are the No. 1 asset within the corporation, and they interact with the customers, who are the No. 1 asset outside the corporation.

The process design organization must understand 11 major interfunctional connections to be successful.

1. Connection with the customer.
2. Connection with the supplier base.
3. Connection with the product design team.
4. Connection with procurement and the material organization.
5. Connection with the systems organization.
6. Connection with marketing and sales.
7. Connection with the organization development portion of the business.
8. Connection with operations.
9. Connection with service.

10. Connection with the financial organization.
11. Connection with the legal organization.

Each of these will be briefly discussed.

*Connections with Customers.*   In the past, the primary contact with the customer has been through marketing and sales. Marketing and sales will continue to have the primary focus. However, as we begin to develop new ways of developing products and rapidly responding to changing customer demands, as Wheelwright mentioned earlier, new manufacturing technologies and approaches must be introduced and developed much more rapidly than they have in the past. The time it takes to respond to a changing customer demand may want to be negative. The enterprise may have to anticipate customer's demands and be ready to go when the customer recognizes that he or she needs that product or service. This goes beyond the notion of "customer satisfaction" to the notion of "delighting the customer." This connection with the customer set is fundamental to the success of the overall enterprise, and process design will play a major role in this integration.

*Integration with the Supplier.*   Fully 50 to 60 percent of many companies' costs of goods sold are acquired from outside organizations. The integration of the process design team with the supplier management team is fundamental to recognizing what is available on the outside, both in terms of equipment availability and material and raw stock availability. Equipment availability is necessary to develop the next steps to utilize the best equipment available throughout the world. Material availability is necessary to develop the next generation products.

*Connections with the Product Design Team.*   Product and process design are now becoming equals in the overall recognition that time-to-market compression is fundamental. The product design team heretofore has been king in many organizations, at least in the Americas. In the Asian Rim, however, the tool engineer has often led the way. Hence, different professional strata have occurred. We will discuss these cultural issues later in this chapter.

The product design team develops the form, fit, and function perceived to be required and, in parallel with the process design team, develops the overall notion of how the product can be developed to meet the customer demands and also to be produced in the enterprise.

**Connection with Materials and Procurement Organization.** The acquisition of materials is fundamental to the overall ability to produce the product. Materials and procurement will have a dramatic effect on the overall process design orchestration. The intriguing thing about this is that these two organizations often use different fundamental elements by which they develop their overall methodologies. Process design often delves into the geometry and the functionality of the product, whereas materials and procurement often deal with bill of material and raw material stock. These two organizations are finding they must come together in an integrated fashion to be of great service to the enterprise.

**Information Systems Connection with Process Design.** Several integrated information systems are necessary to support the overall integration of the process design arena. The systems organizations within many enterprises are now finding that they must represent these users and understand how to absorb and affect and directly and dramatically translate true user requirements into functional systems in a timely manner. Also, the systems world is changing significantly as it relates to users being able to develop systems on their own, given availability of a good data structure with good configuration management, good security, and good, rigorous structure to the system.

**Connections with Marketing and Sales.** The intriguing thing about process design discussing customer requirements with marketing and sales is that these must be formally documented in such a manner that a consistent translation of what the customer wants can be documented and then followed through in a rigorous fashion into the product and process capabilities. The methodology that we have introduced, called quality function deployment, and the follow-through of the true customer requirements into the integrated information systems necessary to support process design and development, have been mentioned many times in this text.

***Connections with the Organization Development Element of the Enterprise.*** With people truly being the No. 1 asset within any corporation, the integration of the process design team with the organization development team is fundamental for the overall nurturing of the multidisciplined employee. Organization development is more than simply looking at the personnel aspects and setting up standard procedures for how people issues will be handled. The organization development responsibility for this enterprise of the 21st century is truly looking at how one can cultivate the capabilities of every employee within the organization. This is a much broader scope than has been addressed in the past, and because of the changing cultural issues that will be addressed in Section 12.2, the organization development team has a major role to play for the overall success of the enterprise.

***Connections with the Operations Activities of the Enterprise.*** Process design and development are only as good as the execution of their results on the factory floor. For this integration to be successful, the teams from process design and operation and management must be fully integrated to understand the current capabilities and the changing capabilities on any factory floor environment at any point in time.

***The Service Connection with Process Design.*** In the past, the process has been primarily viewed as an intermediate step in the life cycle of a product. The process focus had traditionally been looked at as having the first unit out be a useful product. However, as we begin to recognize the true costs in the life cycle, we realize the operation and service cost may be dramatically greater than the fabrication and assembly cost. The customers in the 21st century are looking for reliable products. The reliability, maintainability, and serviceability issues represented by our service organizations must be integrated into the overall process design in a manner such that the product meets not only the initial needs of the customer but the lifetime needs as well.

***The Financial Implications of Process Design and the Integration of the Financial Organization with the Process Design Organization.*** Finance in the past has often been looked on as an organization that does "after-the-fact" accounting for what the enterprise has done. But in the future, proactivity is required. As we've noticed, 70 to 80 percent of the overall product cost is committed during the early stages of conceptual design. Conceptual design often is product and process being developed in a sketch mode with teams of people gathering around a table. This fundamental implication of early financial information to support many fundamental decisions in the product-process development cycle is leading us to recognize that the process design and development connection with finance is fundamental to the overall success of the enterprise. Proactive predictors of financial implications are necessary to compete, and these must be couched in a manner such that the process design engineers and the product design engineers can understand the implications from a cost standpoint early in the product development cycle.

*Interconnection with the Legal Organization.* Certainly with the litigation that exists today, full documentation of how the process is performed and the understanding of the risk management responsibilities for the process are required. Here, the integration with the legal organization becomes fundamental to truly understand how best to document each of the processes performed to develop the overall product and provide that product and/or service to the customer. The connection with the legal organization now becomes fundamental to this overall success.

## SECTION 12.2   THE CULTURAL IMPLICATIONS OF INTEGRATED PROCESS DESIGN AND DEVELOPMENT

As noted in the previous section, process design is becoming a major integral hub of the overall success of the enterprise. It is not a predominant or a subordinate player in the overall product-process development cycle. It is played as a peer and sits at the concurrent engineering roundtable as an equal to product design, materials and procurement, supplier-based management, marketing and sales, and so on.

For this to be successful, fundamental cultural changes appear to be necessary in most companies. The intriguing notion is that we're headed toward what Dr. Joseph Harrington, Jr., called "nondepartmentalized decision making."[2] He states that to be successful in the overall process development cycle, we must integrate the functionalities recognized as distinct elements in today's organization into an organized holistic approach that takes a complete systems view of the entire product-process. Dr. Harrington points out we must break down the barriers of the organizational elements, and we must reduce the notion of professional stratification that exists in many organizations today. Professional stratification is defined as looking on sectors within the organization as being either superior or subordinate to one another.

For example, few companies in the United States have viewed the product design engineer as king, or they have viewed the marketing and sales organization as king, and everyone else would be subordinate to that particular function. It is now becoming obvious that there are no kings in this integrated organization, but significant peer relationships must be established. To do this, Charles Savage points out that we must develop a human network.[3] And that human network takes into consideration all the primary linkages recognized in Section 12.1. We are developing what are called cross-functional, virtual task teams throughout the entire world to develop products and processes in a much more rigorous fashion.

The Manufacturing Studies Board of the National Research Council has recognized the cultural ramifications of integration. A study of 16 successful advanced manufacturing technology installations identified remarkable cultural changes. The results of this study are shown in Table 12–1. The study concluded that fundamental shifts occurring with successful installations of advanced manufacturing technology require significant corporate culture changes. Authority moves to being based on knowledge. Decision making moves to where the action is required, and rather than not training individuals because of their ability to move to competitors, many organizations recognize that they must enhance the employee contributions. Information is shared widely throughout the entire organization. As Dr. Harrington points out, we become nondepartmentalized in the management of information.

The notion that individuals must be rewarded must be balanced with the notion that teamwork is absolutely required. We must recognize that people must be rewarded based on their individual

**TABLE 12–1**
**Culture Shifts in Successful Installations of Advanced Manufacturing Technology**

| Organizational Aspect | Traditional Practice | Shift in Practice |
|---|---|---|
| Authority | Based on position | Based on knowledge |
| Decision making | Close to top | Where action is |
| Employee contribution | Limit knowledge and skills | Enhance |
| Information | Closely control | Share widely |
| Rewards | Individual performance | Teamwork |
| Status | Highlight differences | Mute differences |
| Supervision | *Watchdog* | *Resource* |

Source: MSB Report of National Research Council

contribution, yet encourage teamwork in every aspect of the business. Status will be muted. The bottom line as recognized by the Manufacturing Studies Board is that supervision will change from being a *watchdog* with a span of control of the traditional seven people, which was identified by Frederick Taylor many years ago, to becoming a *resource* with organizational elements of 10, 20, or 30 people under their supporting control. Being a resource is a much different cultural scenario than being a watchdog.

By being a resource, supervisors must openly discuss the issues being addressed, they must directly support the development of the talent and resource base for the process design and all other organizational elements, and they must have access to and authority to acquire all resources necessary to make that team of individuals successful. The reward structure was emphasized by Dr. Ann Majchrzak at a Society of Manufacturing Engineers roundtable discussion held at Arizona State University in December 1990.[4] She stated that before a teamwork reward structure could be fulfilled, all elements of the matrix depicted in Figure 12–1 must be understood and developed.

These notions of rewarding teamwork, of supervisors becoming resources, that information is shared widely, that decision making is moved to where the action is required, and that authority

**FIGURE 12-1**
**The Teamwork Reward Structure Matrix**

|  | Individual | Team | Organization |
|---|---|---|---|
| Financial |  |  |  |
| Non-financial |  |  |  |

*To successfully reward teamwork, research has found that all six of the matrix elements must be clearly defined.*

shifts to being based on knowledge are significant culture shifts that must be addressed as we migrate into the 21st century and view process design and development as a major, integral link to the overall success of the enterprise.

## SECTION 12.3 THE EFFECTS OF INTEGRATED PROCESS DESIGN AND DEVELOPMENT ON SALLY, MARY, DAN, AND BILL

To stress that the major impact of all the integration referenced in this book hinges on the people aspects of the business, we conclude with a case study of the impact on four highly regarded professionals in your company:

- Sally is the well-qualified marketing and sales representative.
- Mary is the fully qualified, degreed design engineer.
- Dan is the manager of purchasing.
- Bill is the fully qualified, degreed manufacturing engineer.

Quite a team! It is one that can get the enterprise to world-class status. What are the implications of this book's issues on Sally, Mary, Dan, and Bill?

Let's begin with Sally. Sally's task is to represent the customer. She has done this for the past several years. But with the notion of integration also comes the notion of "rigor," and Sally is not excited about rigor. Sally says the customers constantly change the specification, go through "hills and valleys" with respect to orders, and love to see Sally compete to get their business. But for the integrated enterprise to succeed, Sally's world changes. First, the application of QFD techniques will allow Sally to represent the customer but in an orderly and rigorous manner. She must begin to quantify the customer's priorities and must understand how to make that customer successful. She must also change the customer's perception of competition to one of strategic alliances. Her objective is to have the customer view Sally's company as a partner that can meet the challenges imposed by the customer and do the job better than anyone else.

Next there is Mary, the design engineer. Mary is a noted conceptualizer. She loves her freedom and creativity. She is the genius behind the successful conceptualization of the past products. Mary's genius is perceived as a core competency of the company. Yet this notion of a concurrent engineering roundtable disturbs her. Mary is an individualist—measured on the number of novel designs she can create. But because of this notion called integration, she is asked to equalize her professional strata with all others around the table—to become a peer. Mary believes this will do two things:

1. Slow her down.
2. Limit her creativity.

Mary's creative genius must be nurtured, but the challenge of management is to establish an environment that will reward Mary not only for her creativity but also for her teamwork. Using the QFD approach and the continuous improvement approach, Mary must learn from the customer and learn from previous projects.

Third comes Dan, the purchasing expert. Dan's task is to look at the supplier side of the enterprise just as Sally looks at the customer side. Dan must join the roundtable with full awareness of the supplier's capabilities just as he would know the capabilities of his own organization. He must also know what is on the horizon at the supplier that he can then nurture up the technological food chain and add value from his company. His task is one of integration, rigor, and strategic alliances.

Finally, there is Bill. Bill always has been asked to join the product development team "too late" in his eyes. He never has had the latitude to contribute in the conceptual phase of the product design because he was never invited. But with this notion of integration comes responsibility as well. There is an old Chinese proverb that says: "Don't ask the question if you don't want the answer!" Bill is now being invited to the roundtable, but he finds that his knowledge bank is weak when dealing with conceptual sketches of the product. Bill realizes the wealth of process knowledge that he and his colleagues have must be restructured to support this new environment. Bill views this as a great opportunity but a great challenge.

Note that Sally, Mary, Dan, and Bill are all hypothetical characters. But maybe they are present in a company you know. The closing assignment for the reader is to establish a strategy and a knowledge worker environment where all of the individuals can flourish with the integration needed for success.

## REFERENCES AND RECOMMENDED READINGS

1. Steven Wheelwright, "Restoring the Competitive Edge in U.S. Manufacturing," California Management Review, Los Angeles, Calif., 1985.
2. Joseph Harrington, Jr., *Computer Integrated Manufacturing,* Krieger Press, Melbourne, Fla., 1973.
3. Charles Savage, *Fifth Generation Management,* Digital Press, Boston, Mass., 1990.
4. Kimberley Beaumariage and Dan Shunk, "Issues in Migrating to Teamwork," CASA Blue Book Series, Society of Manufacturing Engineers, 1991.

# OTHER TITLES IN THE BUSINESS ONE IRWIN/APICS LIBRARY OF INTEGRATIVE RESOURCE MANAGEMENT

## Effective Product Design and Development
## How to Cut Lead Time and Increase Customer Satisfaction
Stephen R. Rosenthal

Shows manufacturing professionals how to shorten the cycle of new product design and development and turn time into a strategic competitive advantage. Rosenthal helps you use an integrated managerial view of the design and development process, which helps companies catch design flaws early, correct mistakes, and avoid long development delays.

ISBN: 1-55623-603-4    $42.50

## Marketing for the Manufacturer
J. Paul Peter

Explains the marketing role to the nonmarketing specialist. Peter provides a detailed analysis of how marketing fits into various organizational structures and product management systems. He offers methods for researching consumer markets and creating a dynamic strategic plan.

ISBN: 1-55623-648-4    $42.50

## Integrated Production and Inventory Management
## Revitalizing the Manufacturing Enterprise
Thomas E. Vollmann, D. Clay Whybark, and William L. Berry

Explains the use of modern planning and control systems to remove excess inventory investment, save on production and distribution costs, and better integrate organization efforts. The authors give you strategies and systems to optimize customer service through the use of the latest inventory monitoring procedures.

ISBN: 1-55623-604-2    $42.50

## Managing for Quality
## Integrating Quality and Business Strategy
Howard S. Gitlow

Details how to integrate quality into the heart of a company's business plan and use it to gain a strategic edge over the competition. Gitlow shows how to satisfy customer requirements and reduce the cost of quality. He offers methods for detecting and preventing defects by using a system that minimizes monitoring efforts.

ISBN: 1-55623-544-5    $42.50

**Managing Human Resources
Integrating People and Business Strategy**
Lloyd S. Baird

Details specific steps for integrating the human resource function into all other key areas, including customer relations and field service. Baird shows how the entire organization is affected by the actions of each of its workers. He offers strategies for achieving human resource goals at the line management level.

ISBN: 1-55623-543-7   $42.50


*Prices quoted are in U.S. currency and are subject to change without notice.*

*Available in fine bookstores and libraries everywhere.*

# INDEX

## A

Absolute control system, 133
Activity, 106–7
Adaptive controllers, 138
Advanced or automated programming tool (APT), 125–26, 134–35
Air Force Manufacturing Technology Division, 125, 126
Allen Bradley, 224–25
Allied Signal Corporation, Garrett Engine Division, 173–80
Alting, Leo, 98, 114
American National Standards Institute (ANSI), 97
American Production and Inventory Control Society (APICS), 239
American Society of Mechanical Engineers, 233
Appleton, Daniel, 181, 210
APT language (advanced or automated programming tool), 125–26, 134–35
Arizona State University, 158
Artificial intelligence-based planners, 104–5
Asset management, 224
Assignment, 123–24
Automated guided vehicles (AGVs), 64–65
Automated material handling, 142
Automated storage and retrieval systems (AS/RSs), 64–65, 142, 143
Automation, 43

## B

Bailey, James, 30, 68, 83, 124, 143
Baird, Lloyd S., 251
Battles, Sean, 180
Beaumarriage, Kimberley, 82
Bedworth, David D., 30, 68, 83, 124, 143
Benchmarking, 43, 61–63, 85–89, 164–65, 172
Berry, William L., 250
Biekert, Russell, 180
Bill of material, 24
Black, J. T., 83
Boeing, 5
Boothroyd, Geoffrey, 50, 83
Bossert, James L., 82, 114
Brown, C. S., 51, 52, 83
Budnick, Frank S., 83, 106, 115
Burbidge, John L., 30, 54–55, 83
Burner, Larry, 56

## C

CAD (computer–aided design); *see* Computer aided design
CADCAM (computer–aided design, computer–aided manufacture), 137
Cahill, Jerry, 83
Camp, Robert, 61–63, 83, 114, 180
Capacity, 28
Cincinnati Milacron, 14
CIRM, 239, 244
Classification and coding mechanism of group technology, 55
Closed–loop control system, 119
Closed–loop feedback control system 119–20
Closed–loop feedforward feedback system, 48–49
Commodity code, 55
Competitive advantage, 146–48

**253**

Complementarity, 10
Composite materials, 39–40
Computer–aided design (CAD), 97, 137, 185–87
Computer–aided manufacturing (CAM), 137
Computer Aided Manufacturing International (CAM-1), 103, 142
Computer–aided program planning (CAPP), 103–5
Computer–integrated manufacturing (CIM), 11
  obstacles to, 14
Computer integrated planning (CIP), 182–84
  requirements, 190–93
Computer numerical control (CNC), 136–38, 140–41, 155
Concurrent engineering, 3, 43, 44–51
  barriers to, 158–62
  closed-loop feedforward feedback system, 48–49
  data distribution, 216–17
  directions and trends, 213–18
  distribution of authority, 50
  quality function deployment, 44–48
  study at Allied Signal Corporation, 173–80
  team formation, 49–51
  user interface, 217–18
*Concurrent Engineering* (Nevins and Whitney), 19
Consolidation, 38
Continuous improvement, 24, 47–48, 56–57, 111–14, 146–147, 149, 212
Continuous path numerically controlled machines, 133
Control engineering, 117–18
Control system, 117
  automatic feedback, 120–23
  closed–loop, 119, 138–39
  definition, 119
  flow of product through factory, 123–24
  information, 193–203
  inventory and material flow, 139–40

Control system—*Cont.*
  machine control, 126–29, 131–33
  migration to integrated process, 140–43
  numerical; *see* Numerical control
  open–loop, 119
  types of systems, 133–34
Core competencies, 5, 61, 146, 148, 211–12
Cost of goods sold (CGS), 203–6
Cost of Quality concept, 114
Critical path method (CPM), 77
  process planning, 106–7
Cross–functional team, 88–89
Customer requirements
  process, 6–9
  quality function deployment, 46–48
Cycle time, 27

# D

Davis, George, 180
Deere Company, 49
Demand–pull, 57–60, 139
Deming, W. Edwards, 43, 82, 211, 212
Dertouzos, Michael L., 238
Design of Experiment (DOE) technique, 114
*Design of the Factory With a Future* (Black), 65
Dewhurst, P., 50, 83
Digital Equipment Corporation, 43, 153, 213
Direct numerical control (DNC), 136–38, 140–41
Dorf, Richard, 119, 143
Downtime, 26
Drees, Jan, 94, 95
Drucker, Peter, 1, 2, 17, 85, 114, 145, 150, 179, 211, 229
Dynamically migratable systems (DMS), 206–8
Dynamic random access memories (Dram), 231

## E

Economic order quantity (EOQ), 10
   defined, 25
   learning curve notion, 155
   setup time, 28, 67
Electronic data interchange (EDI), 213–14
Electronic part manufacturing process, 29, 41
Elimination, 38
Elsayed, Elsayed, 115
Engineering, 117
Engineering bill of material, 24
Equipment, definition, 22
Equipment supplier integration, 221–22
Ettlie, John, 157, 180, 210
Event, 106–7

## F

Factory management and control system (FMCS), 197–204
Factory of the future concept, 188
Feedback control system, 119
   automatic, 120–23
   history, 120–21
   modern examples, 121–23
*Fifth Generation Management* (Savage), 11, 50
Financial systems integration, 203
Flexible composites center, 40
Flexible manufacturing systems (FMS), 90–92, 140–41
   Japanese compared to U.S., 141
Floating zero, 133
Focused factories, 9, 51, 56, 226–27

## G

Galvin, Robert, Jr., 2, 17
General Electric, 47–48
Generative planner, 104
Generic business strategies, 146–48

Geometry profile, 134–36
Giddings and Lewis, 141
Gitlow, Howard S., 250
Globalization with localization, 227
Globerson, Shlomo, 156
Goal space modeling, 223
Goldhar, Joel, 18
Goldratt, Eli, 193, 210
Greene, A., 199, 210
Groover, Mikell, 127, 143
Group technology (GT), 38, 43, 51–56
   assignment and sequencing, 124
   identification of similarities, 51–54
   mechanics of, 54–56
   world class competition, 170–71
Gunn, Thomas G., 210

## H

Hamel, G., 5, 17, 83
Harrington, Joseph J., 11, 12, 18, 125, 245
Hidden cost model, 179
Hitachi, 230
Houtzeel, Alex, 51–52, 83
Hsiang, Thomas, 115
Huang, Philip, 238
Huthwaite, Bart, 50, 83
Hypertext, 216

## I

ICAM (integrated computer–aided manufacturing), 97
*I CAM Def*inition language (IDEF), 70–79
IDEFO/td methodology, 213
Idle time, 26
IGES (standard for simple product definition), 97
Incremental control system, 133
Indirect cost, 26
Indirect material cost, 26

Information management, 2, 181–210
  computer integrated planning, 182–83, 190–93
  computer integrated process system, 208–10
  connections with process design, 242
  dynamically migratable systems, 206–8
  process control systems, 193–97
  requirements, 197–203
    financial systems, 203–6
Integrated information support system (IISS), 208–10
Integrated process design and development, 3, 4–5
  architecture of systems, 187–90
  asset management, 224–25
  benchmarking, 61–63
  concurrent engineering; see Concurrent engineering
  core competencies, 5
  cultural implications, 244–47
  design philosophy, 69
  group technology, 43, 51–56
  information system requirements, 181–210
  interfunctional connections, 239–49
  material handling and flow systems, 63–65
  material management, 57–60
  methodology, 70–76
  obstacles, 2, 14–15
  planning process for, 85–114
  process capability design, 60–61
  simulation, 77–82
  studies
    National Science Foundation, 157–66
    world class manufacturers, 167–73
  total quality management, 56–57
Integrated product–process development (IPPD) (concurrent engineering), 3, 4–5
Integration, 43

Interfunctional connections, 239–49
  customers, 241
  financial implications, 243–44
  information systems, 242
  legal organizations, 244
  marketing and sales, 242
  materials and procurement organizations, 242
  operations activities, 243
  organization development, 243
  product design team, 241–42
  service connection, 243
  supplier, 241
International Standards Organization (ISO), 97
Inventory
  control, 138–39
  as liability, 66
Inventory of knowledge, 151

## J

Jaikumar, R., 140, 141
Japan
  flexible manufacturing systems, 141
  integrated process design and development, 229, 233–38
  prototypes, 231–32
  teamwork, 235–36
Jelinek, Mariann, 18
Jikigata-Ken, 236–38
Just–in–time, 38, 43, 59

## K

Kanban, 38, 43, 59
Kaplan, Robert, 204, 209, 210
Kearney and Trecker, 141
Kibbey, Donald, 129, 143
Kimura, Joseph, 229
Kriegler, Arnold, 33

## L

Lead time, 27
Learning curve, 154–55

LeClair, Steven, 223, 227
Lee, Seouk Joo, 227
Life cycle of products
   cost, 3–4
   traditional, 3
Long–term sustainable competitive advantage (LTSCA), 7–8
Loss of Function concept, 114

## M

McCarron, S., 210
Machine control concepts, 126–33
   classification, 133–34
Machining centers (MC), 141
Macros, 137
Majchrzak, Ann, 246, 247
Make–buy decisions, 105–6
Malcolm Baldrige Award, 108
Manufacturing
   accounting systems, 1
   cycle time, 152–53
   measuring efficiency and effectiveness, 152–53
   types of, 29
Manufacturing resource planning (MRPII), 128–29, 197, 199
Manufacturing throughput time, 27
Master bill of material, 24
Material flow systems, 63–65
Material handling, 63–65
Material management, 43, 57–60
Materials requirements planning (MRPI), 128–29, 199
Material supplier integration, 219–21
Mean time it takes to respond to changing customer demand (MTTRCD) metric, 150–51
Merchant, Eugene, 3, 39, 92
Metallic part manufacturing process, 29, 32–39
Method, definition, 23
Metric; see Process metrics
Modular manufacturing, 1, 85
Mojena, Richard, 83, 106
Monocodes, 55
Monte Carlo simulation, 80, 81, 107–8
Moore, Harry, 129, 143
Motorola, 2
   total quality management, 108
Multimedia presentations, 215
Multivariable control system, 120

## N

National Institute of Standards and Technology (NIST), 70
National Research Council, 158
National Science Foundation studies, 157–66
NCR, 49
Negative feedback control, 121, 122
Nevins, James L., 19, 44, 82
Niebel, Benjamin, 103, 114
Nippondenzo plant works, study, 233–38
Nonmetallic part manufacturing process, 29, 39–41
Numerical control
   advanced developments, 136–38
   classification of machines, 133–34
   closed–loop, 138–39
   geometry profile, 134–36
   history, 125–26
   machine control concepts, 126–34
   principles of operation, 129–33

## O

Object oriented programming (OOP), 207–8
Ocular group technology, 54
Order bill of materials, 24
Orlicky, Joseph, 197
O'Rourke, Tracy, 224–25

## P

"Passing the chalk," 216
Peck, David, 210, 227

PERT (performance evaluation and review technique), 77
   process planning, 106–7
Peter, J. Paul, 250
Plant layout, 24
Plossl, George, 197
Point to point numerically controlled machines, 133
Polycodes, 55
Porter, Michael, 61, 83, 146, 147, 155, 211
Prahalad, C. K., 5, 9, 17, 61, 83, 146, 156, 211
Process, definition, 21–22
Process capability design, 43, 60–61
Process chart, 24
Process control, 117–43
   closed loop, 138–39
   directions and trends, 223–24
   example of modern systems, 121–23
   flow of product through factory, 123–24
   history of automatic control, 120
   inventory and material flow, 139–40
   machine control concepts, 126–34
   migration to integrated process, 140–43
   numerical control, 136–38
   types of systems, 133
Process design and development; *see* Integrated process design and development
Process metrics, 25–28, 145–55
   learning curve, 154–55
   manufacturing efficiency and effectiveness, 152–53
   Nippondenzo Company, 237
   traditional metrics compared, 145–52
Process planning, 22–23, 85–114
   benchmarking, 87–89
   computer aided, 103–5
   control; *see* Process control

Process planning—*Cont.*
   with CPM, PERT and simulation, 106–8
   creation of, 97–103
   defining process needs, 89–94
   directions and trends, 222–23
   make–buy decisions, 105–6
   prototyping, 94–97
   technology assessment, 85–87
   total quality management, 108–14
Process time (PT), 100
Product definition exchange specification (PDES), 97
Product definition exchange standard (PDES), 158
Product definition standards, 96–97
Production flow analysis, 54–55
Production plan, 23
Product process life cycle, 9, 25–28
Product–process development team, 240
Product throughput time, 27
Proof of concept phase in idea creation, 94, 96
Prototypes, 8–9, 94–97, 225–26
   definition, 96
   Japanese, 231–32
Putnam, George, 125

## Q–R

Qualitative process automation (QPA), 224
Quality function deployment (QFD), 8, 20, 44–48
RA 90 storage module, 43
Raw materials, 21
"Real time," 199
Robotics, 142
Rockwell International, 32, 33, 53, 54, 56
Rosenthal, Stephen R., 250
Roth, A. J., Jr., 42, 114, 210
Routers, 128–29
Run quantity, 28

## S

Sakurai, Michiharu, 238
Savage, Charles, 11, 12, 18, 50, 83, 215, 245
Scheduling, 23
Schreffler, Roger, 156
Self-direction, 223
Sequencing, 25, 123-24
Setup time, 28, 66-68
Shingo, Shigeo, 28, 30, 38, 58, 66, 67, 68, 83
Shunk, Dan L., 82, 114, 210, 227
Similarities, group technology, 51-54
Simplification, 38, 43
Simulation, 44
  process planning, 107-8
  role in integrated process design and development, 77-82
Single minute exchange of dies (SMED), 28, 38, 66, 68, 124
Six sigma concept, 10, 108-12
Skinner, Wickham, 9, 17, 51, 55, 83, 211, 227
Slack, 106-7
Smart pallets, 65
SMED; *see* Single minute change of dies
Society of Manufacturing Engineer's LEAD Award, 221, 224
SPC; *see* Statistical process control
Spine concept, 63-64
Stalk, George, Jr., 9, 17, 146, 156, 212, 231
Standard binary code, 130-31
Standard for product definition (IGES), 97
State space modeling, 224
Statistical process control (SPC), 113-14, 219, 221
Statistical quality control, 1, 85
Stauffer, Robert, 42
STEP, 97
Stick figure representations, 185-86
Strategic alliances, 146-47, 148-49, 212

Strategic alliances—*Cont.*
  supplier base, 219-21
Sullivan, William, 83
Supply-push, 57-59
System, 117
Systems approach to management, 2
System integration, 43

## T

Taguchi, Genichi, 113, 114, 115
Taylor, Alex, III, 18, 156
Taylor, Frederick, 19, 246
Teamwork
  concurrent engineering, 49-51, 214-15
  cross-function, 245
  Japan, 235-36
  rewarding, 246-47
  strategic alliances, 149
  world class competition, 170-71
Technical data interchange (TDI), 214
Technological food chain, 221
Technology assessment, 85-87
Throughput time, 26-27
Time, 2, 33, 39
  inventory, 66
  management, 212
  measurement unit, 229
  process plan, 100
  "real," 199
Time-based competition, 146-47, 148, 150-51
Time-to-market compression, 150-51, 232
Tolerance, 28
Tompkins, James, 63, 83
Tools, 22
Toshiba, 231
Total quality management (TQM), 43, 56-57, 146-47
  direction and trends, 218-19
  process design, 108-14
Transducer, 132
Turner, Ed, 56

## U–V

University of Michigan, 158
Value–added function, 21–22
Value–added velocity, 153
Vanderbilt University, 158
Variant planner, 104–5
Visual inspection, 54
Vollman, Thomas, 83, 106, 250

## W

Watt, James, 120
Wheelwright, Steven, 7, 18, 239
Whitney, Daniel, 19, 29, 44, 82
Whybark, D. Clay, 250
Wight, Oliver, 197
Work center, 22
Work in process (WIP), 28, 143
Work measurement, 25

## X–Z

Xerox, 164
Young, John, 229, 238
Zero defects, 114
Zhang, Hongchao, 98, 114
Zimmer, Emory, 127, 143